T0251001

Routledge Revivals

Supply and Costs in the U.S. Petroleum Industry

Any discussion of the various facets of petroleum policy in the United States rests to a greater or less extent on the issue of sensitivity of petroleum exploration, and hence of new petroleum discoveries to economic incentives. Indeed, a principle argument in favour of having a special petroleum policy at all is that domestic petroleum exploration is so sensitive to economic considerations that in the absence of special incentives exploration expenditures would sharply decrease, as would the amount of petroleum discovered; consequently, the nation's known oil resources would be reduced to an extent dangerous in the event of an international crisis.

This study attempts to answer the question: how sensitive are new petroleum discoveries to economic incentives? This book will be of interest to students of environmental studies.

Supply and Costs in the U.S. Petroleum Industry

Two Econometric Studies

Franklin M. Fisher

First published in 1964
by Resources for the Future, Inc.

This edition first published in 2015 by Routledge
2 Park Square, Milton Park, Abingdon, Oxon, OX14 4RN
and by Routledge
711 Third Avenue, New York, NY 10017

Routledge is an imprint of the Taylor & Francis Group, an informa business

© Resources for the Future

The right of Franklin M. Fisher to be identified as author of this work has been asserted
by him in accordance with sections 77 and 78 of the Copyright, Designs and Patents Act
1988.

All rights reserved. No part of this book may be reprinted or reproduced or utilised in
any form or by any electronic, mechanical, or other means, now known or hereafter
invented, including photocopying and recording, or in any information storage or
retrieval system, without permission in writing from the publishers.

Publisher's Note
The publisher has gone to great lengths to ensure the quality of this reprint but points
out that some imperfections in the original copies may be apparent.

Disclaimer
The publisher has made every effort to trace copyright holders and welcomes
correspondence from those they have been unable to contact.

A Library of Congress record exists under LC control number: 77086394

ISBN 13: 978-1-138-88724-4 (hbk)
ISBN 13: 978-1-315-71420-2 (ebk)

SUPPLY AND COSTS IN THE U.S. PETROLEUM INDUSTRY

Two Econometric Studies

I The Supply Curves of Wildcat Drilling and of New Petroleum Discoveries in the United States

II Measuring the Effects of Depth and Technological Change on Drilling Costs

By Franklin M. Fisher

RESOURCES FOR THE FUTURE, INC.

Distributed by The Johns Hopkins Press, Baltimore, Maryland 21218

© 1964 by *Resources for the Future, Inc., Washington, D.C.*
Library of Congress Catalogue Card No. 64–25067
Price $5.00

RESOURCES FOR THE FUTURE, INC.
1755 Massachusetts Avenue, N.W., Washington, D.C. 20036

Board of Directors Reuben G. Gustavson, *Chairman,* Horace M. Albright, Erwin D. Canham, E. J. Condon *(Honorary)*, Joseph L. Fisher, Luther H. Foster, Hugh L. Keenleyside, Otto H. Liebers, Leslie A. Miller, Frank Pace, Jr., William S. Paley, Laurance S. Rockefeller, Stanley H. Ruttenberg, John W. Vanderwilt, P. F. Watzek

President, Joseph L. Fisher
Vice President, Irving K. Fox
Secretary-Treasurer, John E. Herbert

Resources for the Future is a non-profit corporation for research and education in the development, conservation, and use of natural resources. It was established in 1952 with the co-operation of the Ford Foundation and its activities since then have been financed by grants from that Foundation. Part of the work of Resources for the Future is carried out by its resident staff, part supported by grants to universities and other non-profit organizations. Unless otherwise stated, interpretations and conclusions in RFF publications are those of the authors; the organization takes responsibility for the selection of significant subjects for study, the competence of the researchers, and their freedom of inquiry.

RFF PUBLICATIONS STAFF: Henry Jarrett, *editor;* Vera W. Dodds, *associate editor;* Nora E. Roots, *assistant editor.*

To my grandparents

Foreword

In these pioneering studies, two important elements in the dynamics of petroleum supply are subjected to rigorous quantitative analysis: the responsiveness of oil discoveries to price incentives, and changes in drilling costs over time. People in industry, government and universities who are concerned with the economic problems of petroleum will find this book of interest not only for the significance of Franklin M. Fisher's findings, but for the analytical techniques he has applied to a set of particularly complex problems. Those who are interested in a further probing of these problems may find Mr. Fisher's approach useful in analyzing statistics covering different periods of time.

It was originally an interest in analyzing U.S. oil import policy which led the author to research on the specific questions covered in the present studies. Finding that existing knowledge was not adequate for performing a rigorous analysis of the economic effects of import policy, Mr. Fisher chose to devote his efforts to the more basic task of studying the conditions of petroleum supply in the United States. The wisdom of this choice is well attested by the fresh insights into the process of petroleum exploration and the behavior of drilling costs which are achieved by his scholarly and imaginative analysis of the underlying statistics, as well as by the solid foundation that he has provided upon which new research may be built.

Mr. Fisher undertook most of the research for these studies while at the Massachusetts Institute of Technology, where he is associate professor of economics. His research was sponsored by Resources for the Future as part of a program of studies covering numerous aspects of the economics of the petroleum industry and its institutional setting. Additional research now under way deals with the measurement of oil and gas reserves; petroleum conservation regulations; and various facets of oil as an internationally traded commodity. We hope that the components of our research program will, over time, lead to an improved understanding of the functioning of the petroleum industry which will give those concerned with its operations a better basis for reaching decisions as to its future course.

Sam H. Schurr
Director, Energy and
Mineral Resources Program
Resources for the Future, Inc.

Preface

The two studies in this volume are presented in the belief that intelligent evaluation of trends or policies affecting the petroleum industry can be substantially aided by econometric investigations. This is true whether such evaluation takes place at the firm, industry, state, or national level.

Unfortunately, despite the many efforts made by the industry, in many areas petroleum statistics are not as reliable or as useful as they might be. This is particularly the case where there is substantial reluctance to describe in any detail the methods whereby particular series were estimated, but it is not restricted to such situations. If, accordingly, I have been critical in some places of the data presented by various industry sources—particularly those of the *Joint Association Survey of Industry Drilling Costs* which I discuss at length in the second study—it is not because I lack appreciation of the difficulties of the tasks involved. Rather, it is because I hope that my criticism will aid in performing those tasks in the future.

These studies were supported by grants from Resources for the Future, Inc., first to the University of Chicago and then to the Massachusetts Institute of Technology. While I am grateful to the organization as a whole, I feel that I owe a special debt of gratitude to Sam H. Schurr, director of its energy and mineral resources program, for his patience and encouragement.

Others who aided me by their willingness to discuss some or all of the problems involved were Robert G. James, Chester Doyle, and other members of the Socony Mobil Oil Company; and Jules Joskow and Bruce C. Netschert, of National Economic Research Associates. I also spoke with, and had aid in securing data from, a large number of other people in the industry and at Resources for the Future whom space forbids my thanking by name. At M.I.T. I had helpful discussions with Albert Ando and Michael Rieber, among others.

The work was revised during my tenure of a National Science Foundation Postdoctoral Fellowship at the Econometric Institute of the Netherlands School of Economics, in Rotterdam. During this revision I had helpful discussions with Marc Nerlove, of Stanford University, and Thomas J. Rothenberg, now of Northwestern University. I received helpful comments on an earlier version from Mr. Netschert, David V. Hudson, Jr., of the Humble Oil and Refining Company, Stephen L. MacDonald of the University of Texas, and Orris C. Herfindahl and Perry D. Teitelbaum of Resources for the Future.

None of these persons or organizations necessarily share my views and I am entirely responsible for any errors.

The bulk of the computations reported in the study of drilling costs was performed at the M.I.T. Computation Center and was programmed by Richard LaBrie. This short statement cannot convey my gratitude for the immense amount of work Mr. LaBrie put into the project over fully two years. The computations reported in the first study were performed at the Computer Facility installation of the School of Industrial Management at M.I.T. Merrill J. Bateman, George Delehanty, Malcolm Gallatin, Arthur Wright, and especially William Oakland and Cynthia Travis provided a large amount of research and computational assistance without which the studies could not have been performed. Beatrice A. Rogers typed the manuscript; Grace C. Locke and Linda Thompson, the tables.

Finally, my wife Ellen cheered me in bad times and rejoiced with me in good. She proofread and copied numbers, patiently listened to me discourse on the problems at length, and helped edit the results. My son Abraham also aided by agreeing to chew up only non-convergent results on drilling costs.

F. M. F.

Cambridge, Massachusetts
                and
Rotterdam, The Netherlands

Contents

List of Tables

I

The Supply Curves
of Wildcat Drilling and of
New Petroleum Discoveries
in the United States

1. Introduction: The Problem

Any discussion of the various facets of petroleum policy in the United States—percentage depletion, import quotas, restriction of production to market demand—rests to a greater or less extent on the issue of the sensitivity of petroleum exploration, and hence of new petroleum discoveries to economic incentives. Indeed, a principal argument in favor of having a special petroleum policy at all is that domestic petroleum exploration is so sensitive to economic considerations that in the absence of special incentives exploration expenditures would sharply decrease, as would the amount of petroleum discovered; consequently, the nation's known oil resources would be reduced to an extent dangerous in the event of an international crisis.[1]

The merits of such an argument clearly turn on a great many points which can only be matters of speculation. Is it true, for example, that an international crisis is likely to be of long duration requiring the provision of large supplies of petroleum over a relatively long period, or is domestic policy geared to the last crisis but one? We shall not attempt to

---

[1] See, for example, the testimony in 1959 before two congressional committees by R. J. Gonzalez, reported in *Percentage Depletion for Petroleum Production* and *Analysis of the Domestic Oil Industry* (Houston: Humble Oil and Refining Co., 1959). Industry spokesmen do not seem to make the argument that very high incentives may be required if exploration and discovery are *insensitive* to economic considerations. Discussions generally turn on whether the removal of an existing incentive will lead to substantial reductions in oil discoveries rather than on whether powerful new incentives are required to increase them. The present analysis is clearly relevant to both situations.

settle such points as this. Yet there is at least one crucial question which is in principle amenable to econometric investigation: the question of the sensitivity of exploration and discovery to economic incentives is surely a quantitative question with a quantitative answer. It is one thing to say that exploration and discovery *are* sensitive and another to say *how* sensitive. Any policy evaluation must begin with some such estimate if the argument under discussion is to be intelligently weighed.

This study presents such an estimate. The results of econometric analysis, which are summarized in the next section, do seem to go a long way toward answering the question: How sensitive are new petroleum discoveries to economic incentives? As an aid in the evaluation of petroleum policy, they should be useful. However, for a variety of reasons they should be regarded as an aid, not as a substitute for policy evaluation.

Why this should be so is inherent partly in the nature of the data and partly in the complex character of the petroleum industry itself. First, the coefficients presented in this study, as in any other study of a similar nature, are important for their orders of magnitude, not for the apparent precision of the numbers themselves. These size relationships, however, reflect positive results that have been subjected to empirical testing. There is ample evidence in this study, for example, that the average size of discovery decreases when price rises. The reasons offered for this behavior may be open to question, but not the fact itself.

Second, it would be desirable, were it possible, to take as the basis for analysis the quantity of petroleum that is readily available, this information being directly related to policy evaluation. But organized statistics on development drilling and its results do not yet exist in a usable form. Consequently, our analysis concentrates on wildcat drilling and on estimates of the total amounts of oil and gas in known fields. It seems likely that qualitatively similar results to those here found would apply to development drilling and its results; this is impossible to verify, however, in the present state of industry statistics.

Third, enough quantitative evidence is provided in the following pages to disprove a commonly held contention that random influences on exploration and discovery are too large to permit any quantitative study of systematic influences on the discovery process.[2] However, more than quantitative evidence is required to say whether petroleum discoveries are insensitive, sensitive, or very sensitive to economic incentives; answers depend also on the context in which the question is posed and the frame of reference of the questioner. The evaluation of particular aspects of petroleum policy

---

[2] On the general question involved in such a criticism, see O. C. Herfindahl, *Copper Costs and Prices, 1870–1957* (Baltimore: Johns Hopkins Press, 1959), Chapter 3. Econometric analysis does leave explicit room for random elements, and the present study produces a good deal of quantitative evidence on the influence of systematic factors on exploration and discovery.

is a complex business to which studies such as the present one are merely a prerequisite.

2. A Summary of Results

Exploration versus Discovery
(See pages 6–8, 23, 35–39)

The principal conclusion reached in this study is that there is an important substantive distinction between the supply curve of exploratory effort on the one hand and the supply curve of new petroleum discoveries on the other. This is the case because economic incentives not only influence the amount of exploration that occurs; they also determine its characteristics. Thus, it is shown that an increase in economic incentives does lead to more wildcat drilling but that this takes place on prospects poorer than those which would be drilled at a lower incentive level. It follows that measures of average results, such as average size of discovery or the success ratio (the ratio of productive to total wildcats), are functions not only of the distribution of petroleum prospects found in nature, but also of the risk attitudes of operators in the industry. Such measures are characteristics of the distribution of prospects *accepted* in any given time period rather than solely of the distribution of prospects considered. Thus, for example, the success ratio cannot be taken as a measure of the probability of discovery, as is frequently done. Since total new discoveries are the product of (1) the number of wildcats drilled, (2) the success ratio, and (3) the average size of discovery per successful wildcat, and since the latter two factors, other things being equal, can be expected to decline with a rise in economic incentives, the response of new discoveries to such incentives will generally be smaller than the response of wildcat drilling itself.

This theoretical argument is amply supported by the results of our empirical analysis which, for reasons of data availability, covers the period 1946–1954–55. Other things being equal, an increase of 1 per cent in the deflated price of crude at the wellhead tended during that period to increase wildcat drilling by roughly 2.85 per cent but also induced a *decline* in the success ratio of about .36 per cent and a *decline* in the average size of discovery of about 2.18 per cent. The net result was thus to increase new discoveries by about .31 per cent, a figure very much smaller than the approximate 2.85 per cent applying to the effect on wildcat drilling itself (although the large size of the difference may well apply principally to relatively short-run direct effects). It is the general order of magnitude of the figures, not the figures themselves, which verifies the conclusion of the theoretical argument just described. The general relationships are maintained even when the analytical procedures are varied, and they are strongly supported by evidence from corresponding analysis for natural gas. It is thus fallacious to use the sensitivity of exploratory effort to economic incentives as a measure for the sensitivity of new discoveries to such incentives. It is the latter sensitivity which is important in evaluating policy, and clearly it is substantially less than the former.

In addition to the principal conclusion just described, a number of other results are obtained which are of some interest for the light they shed on the behavior of exploration and discovery in the domestic petroleum industry. The first of these is directly related to the principal result, as it deals with the effects of an incentive system that is partially operated by maintaining crude prices through restriction of allowable production.

The Effect of Production Restrictions
(See pages 24, 34–35)

In the period under study, restrictions on the daily output of a given oil well were not particularly onerous, being far less stringent than in the period since Suez. Hence, estimates of the effects of the earlier restrictions are likely to be very different from those that would be obtained from current data were such available. This is borne out by the results of the analysis, which show only minor effects (see pages 34–35), and the estimates are not nearly as reliable as are most of the others obtained. Bearing this in mind, the results indicate that the effect of an increase in Texas shut-down days (used as a measure correlated with over-all restrictions) is to shift wild-cat drilling to other parts of the country with less severe restrictions than Texas and to somewhat reduce over-all exploratory activity. It would appear, therefore, that when crude price is maintained by production restrictions, not only is the geographical pattern of exploration slightly distorted, but the procedure is less efficient than direct subsidy as an incentive to exploration, because the positive effects of a high crude price are partly offset by the negative effects of the restrictions themselves.

Size versus Certainty and the "Inventory" of Small Prospects
(See pages 25–27)

Two hypotheses concerning industry behavior were partly suggested by one set of empirical results and tested on another set. They are advanced as somewhat more tentative than is the principal conclusion. The first concerns the shape of the distribution of prospects considered in any one year and the effects of that shape on exploratory activity. The second concerns the role of natural gas, discussed in the next subsection.

For economic as well as possibly for geologic reasons, the small prospects considered by operators tend to be relatively certain and the big prospects relatively risky. This is so if only because big prospects, by offering larger returns on investment, attract operators at higher levels of risk than do small prospects. Over time, an "inventory" of undrilled prospects accumulates about which something is known but which are either too small or too uncertain to be worth drilling under current conditions. These prospects normally may not be considered by top management, but may be kept on file by scouts. Information on new prospects adds to the inventory and further information about the larger known prospects tends to deplete it, leaving a residue consisting largely of small, relatively certain prospects. When the price of crude rises or other conditions appear to warrant increased production, outflow from this inventory tends to increase above normal as smaller prospects become worth drilling.[3] Since small prospects in the inventory tend to be relatively certain, the fraction of wildcats that are productive tends to rise. This may offset the tendency for riskier prospects to become attractive at an increased incentive level. In any case, other things being equal, a rise in the success ratio for wildcats for whatever reason tends to be associated with a depletion of the inventory of relatively small, relatively certain prospects.

This set of circumstances thus tends to be followed by an increase in average discovery size and an increase in risk above the level that would otherwise have obtained. There will also be a decrease, *ceteris paribus*, in wildcat drilling as the set of acceptable prospects will be smaller than would otherwise have been the case. To put it most strongly, the argument on page 3 relating to the principal conclusion showed that a rise in the success ratio is not the same thing as an increase in the probability of discovery; the present argument leads to the conclusion that such a rise tends to be associated with a *decreased* probability of discovery in the following year, other things being equal.

The Role of Natural Gas in the Period Studied
(See pages 18–19, 27–29)

If, as has been suggested, the role of production restrictions in influencing exploration and discovery has probably changed since 1946–1954–55—the most recent period providing data for this analysis—the role of natural gas discoveries and prices has changed even more. Since the end of that period, natural gas has become a product of major importance and its discovery can yield a return comparable to that of crude petroleum discovery. This was far from the case in the period studied, during most of which new discoveries of natural gas were likely to be unmarketable for a rather long time, the demand for natural gas being extremely thin. Several years might have to elapse before the gas from a new field could be brought to market through newly built pipelines.

---

[3] Note how this is related to the argument made for the principal conclusion of the study, already discussed.

The first consequence of this lag in development is that the distant nature of returns on investment in exploration and the uncertainty associated with the future price at which a new gas discovery might some day be sold made the current price of gas relatively unimportant as an incentive to discovery.

Moreover, because the search for oil and gas is carried out jointly, it is not unreasonable to suppose that, given the probability of finding either oil or gas, the higher the ratio of past gas discoveries to past oil discoveries, the lower will be thought the probability of finding oil in a given area, other things being equal. It follows that the discovery of large gas fields may have acted, other things being equal, as a disincentive to exploration and to the drilling of large structures. Clearly, this is no longer the case at the present time.

3. The Many-Faceted Drilling Decision and the Statistical Record of the Industry

The focus of this study is on the aggregate supply curves of wildcat drilling and of new discoveries, rather than on the decisions made by individual operators.[4] It is the case, of course, that aggregate industry curves are made up of the drilling decisions of individual operators, so that the influences we seek to measure will primarily be reflections of forces acting on individual decision makers; nevertheless, no attempt is made here to duplicate the individual drilling decision.

The most important consequence of this is that there are certain items which the individual can take as given but which, as measured for the industry as a whole, are not independent of industry decisions. Thus, consider measures such as the success ratio (the fraction of wildcat wells that are not dry holes) or the average size of fields discovered. It is entirely wrong to suppose that such magnitudes are determined independently of economic incentives in the same sense as is the probability of finding something or the expected size of any find for an individual prospect.[5] The reason for this is that measures such as the success ratio and average size of actual discoveries[6] are not simply measures of properties of the distribution of opportunities presented to the industry in a given year; they are measures of properties of the distribution of opportunities *accepted* by the industry and this is not at all the same thing. Such measures of industry performance

---

[4] This is a quite different focus from that of C. J. Grayson, Jr., *Decisions under Uncertainty: Drilling Decisions by Oil and Gas Operators* (Boston: Harvard University, Division of Research, Graduate School of Business Administration, 1960).

[5] Even the latter numbers—in the aggregate—are not independent of economic incentives, since the set of available prospects depends in part on past drilling decisions and exploratory activity. This is not the same as the problem being considered which involves the relation of current decisions and results to current incentives.

[6] The measurement of this variable is itself quite a problem. See pp. 16–17.

depend both on the opportunities found in nature and presented for possible drilling, *and* on the drilling decisions made. Indeed, they are an essential part of the description of the average drilling decision.

An example will aid us here. Suppose that the price of crude rises somewhat and that the new price is expected to persist. What will be the result? Clearly, the number of wildcats drilled will be higher than would have been the case at the old price. Prospects considered to be attractive before the price change will now offer greater rewards; but, in addition, prospects that hitherto were considered to be marginal will now be sufficiently attractive to justify the expense of drilling them. The increase in wildcat wells, then, will come from the drilling of these additional, poorer prospects. Assuming that operators' forecast as to the relative attractiveness of the prospects are approximately correct, it follows that the statistical record of actual industry performance will show a decline in one or more dimensions: the success ratio may be lower than it would have been at the lower price of crude; fields discovered may be deeper; and, what is most likely, average discoveries may be smaller. Which of these effects or which combination of them will occur will depend both on the actual distribution of prospect attributes and on the risk attitudes of operators. It goes without saying that if the price of crude falls instead of rising, reverse effects take place as the less attractive prospects are cut from the drilling list.

The general point should now be clear. *Even with the same set of prospects, a change in economic incentives induces a change in the record of success and costs in drilling. While that record is indeed largely the result of "natural conditions," it is by no means independent of economic factors.* This conclusion is strongly supported by the empirical results reported below.

It follows that the question of the sensitivity of wildcat drilling to economic incentives and the question of the sensitivity of new oil discoveries to those incentives are not the same thing. They differ because the total of new oil discoveries is the product of three terms: the number of wildcats drilled; the fraction of such wells that are producers; and the average size of field discovered per producing wildcat. The sensitivity of the product depends on the sensitivity of all three factors to economic incentives rather than solely on the sensitivity of the first of them. The responses of size of field and success ratio to economic incentives are indissoluble parts of the question of the response of new discoveries and of wildcat drilling itself. All these matters are but part of the same basic economic decision as to whether and where to drill.

It follows further that a proper analysis of the supply curves of wildcat drilling and discovery cannot proceed by means of the *ex post* industry statistics, since these are based on the proportion of successful strikes and size of fields found out of a given number of wildcat drillings. The decision to drill is indeed likely to be a function of the average probability of success and the average size of field expected before drilling decisions are made, but we have just seen that these are not the same thing as the variables measured by the *ex post* industry statistics. The drilling decisions of the industry start from the characteristics of the distribution of prospects considered and produce *both* the number of wildcats drilled *and* the observed average characteristics of the prospects on which they are drilled. One set of variables does not determine the other; rather, all these things are jointly determined in the course of the drilling decisions taken. At the risk of repetition, let us emphasize this point again. The decision to drill a particular prospect depends on its characteristics and on other factors; the total number of wells drilled does not depend on the average characteristics of the prospects drilled as viewed after the event any more (or any less) than those average characteristics depend on the number of wells drilled. Hence the decisions which determine how many wildcats are drilled are the same as the decisions determining the prospects on which they are drilled. The relationship involved is an interdependent one which cannot be analyzed except as such.

It is logical, therefore, to view the wildcatting decisions made by the industry as a process whose inputs are various natural and economic stimuli and whose outputs are the number of wildcats drilled and the average characteristics of the results of that drilling. In the following section we shall outline the influence of each of the inputs on each of the outputs.

4. The Variables of the Model

What, then, are the inputs and outputs of our model? Presumably the drilling decision is influenced by profit considerations, so that the aggregate model must contain variables which reflect as well as possible the forces which enter, explicitly or implicitly, into calculations of probable returns and costs from the average prospect considered in a given year.

Such reflection must necessarily be inexact. One important reason for this has already been indicated: We cannot from the data available observe the average characteristics of the prospects considered but only certain characteristics of the prospects accepted. This is unfortunate, since it is obvious that a prime determinant of the drilling decision is the nature of the prospects available. Thus it is necessary to use a surrogate here—the characteristics of the prospects drilled in the preceding year. Three of the model's explanatory variables will thus be identical with three of the variables to be explained lagged one year: the success ratio, the number of barrels of oil found per productive wildcat, and the number of million cubic feet of natural gas found per productive wildcat.[7] While this is not a wholly desirable situation, it does avoid the logical problem posed in Section 3: The results of last year's drilling can be known (or estimated in the case of size of discovery) before making this year's decisions in a way in which the results of this year's drilling cannot.

Furthermore, our unit of observation will be taken to be the experience of each year in each of the five petroleum districts. These districts are used rather than some more natural unit because some of the data are broken down only as far as petroleum districts. This is not desirable, but there is nothing that can be done about it in the present state of industry statistics.[8] The three lagged variables in question thus serve as a crude indicator of the information available to wildcatters on the average properties of prospects in the given petroleum district. This is only a rough substitute for the detailed information actually acquired by wildcatters from past experience, but it is all that is available to us.

These three lagged variables serve in another sense to distinguish among petroleum districts. Clearly, different districts differ as to probability of success and expected discovery size. Districts which have had large structures, for example, tend to continue to have them. By using the lagged value of the dependent variable in each of the regressions which are to "explain" those variables, such inter-district scale effects which persist over time are taken out; thus the influence of the other variables in the regression are left more or less free of such effects.[9]

It follows that the three lagged explanatory variables do triple duty. First, they are a rough indication of the characteristics of this year's prospects; second, they crudely measure the information available to wildcatters as to returns in a given area at a given time; third, they distinguish between

---

[7] For simplicity, natural gas was excluded from the discussion of drilling decisions in Section 3. It should be clear that much the same considerations apply to size of gas discoveries, although the influences on it may be somewhat different (see the discussion on pp. 18–19). Natural gas liquids will be ignored as rather small. To the extent that discoveries of natural gas liquids are proportional to discoveries of petroleum (or of natural gas) across petroleum districts, the effect on our results is even more negligible than this would indicate.

[8] The composition of the five petroleum districts is as follows: DISTRICT I: Maine, New Hampshire, Vermont, Massachusetts, Rhode Island, Connecticut, New York, New Jersey, Pennsylvania, Delaware, Maryland, Virginia, West Virginia, North Carolina, South Carolina, Georgia, Florida, and the District of Columbia. DISTRICT II: Ohio, Kentucky, Tennessee, Indiana, Michigan, Illinois, Wisconsin, Minnesota, Iowa, Missouri, Oklahoma, Kansas, Nebraska, South Dakota, and North Dakota. DISTRICT III: Alabama, Mississippi, Louisiana, Arkansas, Texas, and New Mexico. DISTRICT IV: Montana, Wyoming, Colorado, Utah, and Idaho. DISTRICT V: Arizona, California, Nevada, Oregon, Washington, and Alaska.

[9] Another way of describing this procedure is as follows. It seems plausible to assume that the residual disturbance in each equation is composed of two parts: a time component constant over districts and a district component constant over time. In putting the lagged value of the dependent variable into the regression, we are putting in a variable highly correlated with the district component of the disturbance. This biases the estimated coefficient of the lagged variable in question towards unity but removes the effect of the district component of the disturbance on the estimates of the remaining coefficients. (It should also be noted that this procedure biases estimated standard errors downward so that the good fit to data obtained in the results is in some part illusory.) The author is indebted to Marc Nerlove for this intrepretation.

petroleum districts in our regressions. The first of these roles, in particular, must be interpreted with extreme care—it is not the case, for example, that a high success ratio in one year means that the average probability of success in the following year is necessarily high; however, all three must be carefully considered in interpreting the results.[10]

In essentially similar fashion, the average depth of last year's new field wildcats is used as an explanatory variable. It serves as a measure of cost of development, among other things.

This brings us to the more purely economic explanatory variables. The calculation of expected returns and costs from a prospect clearly involves a number of forecasts that cannot be reproduced here. In particular, the proper calculation of costs per barrel of oil and of the present value of such costs is a highly complex operation that cannot possibly be accomplished in the present state of industry statistics.[11] Only a crude allowance for expected costs can therefore be made. This is done by deflating the price variables by (1) the wholesale price index—on the assumption that costs reflect economy-wide trends—and, alternatively, by (2) each of two indices computed from the drilling cost indices of the Independent Petroleum Association of America. Since depth trends are also included in the model and since most of the variance in prices is geographical, this procedure should be adequate.

As for price itself, what is clearly wanted is *expected* price over the life of the field. However, since crude oil prices are so stable,[12] current price seems to be a reasonable indicator of prices expected over at least the next few years. For gas prices, this is not the case, and thus a three-year moving average of prices is used here.

However, consideration of expected returns clearly involves consideration of the time profile of returns and hence of expected allowables. Here, too, it is assumed that current allowables are projected forward.[13] Since the influ-

---

[10] There is, of course, some choice as to *which* past values of the variables in question to use in our analysis. The decision to use only one past year was made for the following reason. As most of the variation in the variables in question is across petroleum districts and as inter-district differences tend to persist over time, last year's observations are highly correlated with those for previous years. It follows that the use of such earlier observations adds little information and that the separate introduction of such observations would not reliably distinguish among the effects of different past years.

[11] See J. E. Hodges and H. B. Steele, *An Investigation of the Problems of Cost Determination for the Discovery, Development, and Production of Liquid Hydrocarbon and Natural Gas Resources*, The Rice Institute Pamphlet (Houston), XLVI, No. 3 (October, 1959); and W. F. Lovejoy and P. T. Homan with C. O. Galvin, "A Study of the Problems of Cost Analysis in the Petroleum Industry," *Journal of the Graduate Research Center* (Southern Methodist University), XXXI, Nos. 1 and 2 (February, 1963). As is made clear in these references, "replacement costs" as used in the industry are an improper measure for our purpose, and it is problematical if they are meaningful at all.

[12] We shall be considering 1946–1955 for the most part, as some of the data end there. This is not the place to comment on the reasons for the stability of crude prices.

[13] This assumption is probably safer for the years for which the estimates were made than for a more recent period. On the projection of allowables by operators, see Grayson, *op. cit.*, pp. 122–25; on prices, pp. 113–19. Note that if prices are projected upward to account for general inflationary trends, our use of a price deflated by the wholesale price index takes care of this.

ence of allowables may well be different in Petroleum District III (which includes Texas) from their effect in the other four petroleum districts, this variable is broken into two variables to account for the difference. (See pages 19–20.)

One other economic variable must be mentioned, namely, the effect of the income tax laws. Since the provisions for percentage depletion did not change over the period examined, there is no direct way of measuring its effect. On the other hand, it is obvious that a cut in the depletion allowance influences the drilling decision by affecting expected profits, and so does a fall in crude prices. It follows that the sensitivities of exploration and discovery to the size of the depletion allowance must be similar to their sensitivities to the price of crude oil.

Finally, just as the variables measuring the results of last year's drilling are used as scale variables which distinguish among petroleum districts in the equations for the corresponding current variables, so it is necessary to have a variable in the equation for the number of wildcats drilled which measures scale by taking account of the number of prospects seriously considered. The measure used for this is geophysical and core drilling crew time employed, since relatively few wildcats are drilled without the previous employment of such crews. As economic incentives may well influence the number of prospects seriously considered, it is clearly necessary to test such effects by attempting to explain this variable. The drilling decisions of the industry are thus viewed in two stages: the first is the decision as to gathering information on prospects; the second is the drilling decision itself which takes place, in some firms, at a higher level of management. The second of these at least is sensitive to economic influences.[14]

In principle, since the factors which influence the second stage of the drilling decision are presumably much the same as those which influence the first (the employment of geophysical and core drilling crews), it is possible to dispense with the crew-time variable by substituting for it in the equations in which it appears. In other words, one might simply short-circuit the two stages and estimate the effects of other variables directly on wildcat drilling. But because data on crew time are available, this would be undesirable for a number of reasons. First, such crew time has an intensive as well as an extensive information-gathering role and thus influences wildcat-drilling and its results in more than one way; putting together the two relationships involved would partly destroy our ability to obtain structural information. Second, such a procedure would seriously hamper the testing of our estimates by putting together the random elements of two relation-

---

[14] Note that there is thus no logical difficulty about using geophysical crew time as an explanatory variable. Because the decision to gather information logically takes place prior to the drilling decision itself, the problems of differentiating between the characteristics of wildcats drilled and those considered to be worthy of drilling (discussed on p. 7) do not arise. In technical language, the model is block recursive with the equation for crew time one block and all other equations a second one. See F. M. Fisher, "On the Cost of Approximate Specification in Simultaneous Equation Estimation," *Econometrica*, 29, No. 2 (April, 1961), pp. 149–52.

ships. This is important here because the results obtained indicate that the influences of various factors on information-gathering activity are likely to be rather different and far less systematic than their effects on wildcat drilling with such activity held constant. Finally, it is important to have inter-district effects in the equation for the number of wildcats drilled accounted for by the crew-time variable as a measure of the scale of opportunities and of activity. In this way, the effects of the other variables involved in the relationship can be measured without large error stemming from the effects of inter-district variation.

To sum up, the variables to be explained by the model are: the number of wildcats drilled; the success ratio; the amount of oil found per successful wildcat; the amount of gas found per successful wildcat;[15] the amount of geophysical and core drilling crew time employed. The principal explanatory variables are: the results of the previous year's drilling as measured by lagged dependent variables; the price of oil; and variables measuring production restrictions.

5. Long-Run versus Short-Run Effects and the Pooling of Observations from Different Petroleum Districts

At this point, it is necessary to discuss in some detail the reasons for and the consequences of the pooling of observations on different petroleum districts in the same regression.

Obviously, an important reason is the desirability of gaining observations in a model with a large number of variables. Were national aggregates used, or each petroleum district analyzed separately, the number of degrees of freedom involved would be very small if not zero. This is so because some of our data are not available after 1954 and because technological change in discovery techniques would make the use of prewar and postwar data in the same regression extremely dubious.[16] On the other hand, the inclusion of additional observations is positively undesirable if such observations were not generated by the mechanism under investigation.[17] It follows that extreme care must be taken in adopting such a procedure and in interpreting the results obtained thereby.

The issue is whether the economic relationships[18] being investigated are such as to be affected by variations among petroleum districts in ways dif-

---

[15] No distinction is made between "gas" and "oil" wildcats, as that classification is arbitrary and entirely a matter of hindsight. Since oil and gas are joint products and the search for one is inseparable from that for the other, they are treated symmetrically.

[16] Such techniques have not changed very much since the war. See B. C. Netschert, *The Future Supply of Oil and Gas* (Baltimore: Johns Hopkins Press, 1958), pp. 45–46.

[17] See F. M. Fisher, *A Priori Information and Time Series Analysis* (Amsterdam: North-Holland, 1962), Chapter 1.

[18] Note that the question is one of relationships and not of the values of the variables themselves.

ferent from effects taking place across time and of which explicit account has not been taken.

On the whole, this seems unlikely to be the case in any important sense. The same forces that affect the drilling decision in different years in a given petroleum district also affect the distribution of drilling decisions over petroleum districts. The amount and type of wildcat drilling are different among districts because petroleum districts differ in the characteristics of the distributions of wildcat drilling opportunities to be found in them and because of differences in the price of the crude oil (and of the natural gas) to be found. Wildcat drilling is less in one district than in another, and occurs on prospects of different size and riskiness, because it is harder to find oil, because fields tend to be smaller, because prices at the wellhead tend to be lower, for some combination of these reasons, and so forth. But these, on the whole, are the very forces that we are including as explicit variables. Thus it is that a primary reason for including the lagged values of the success ratio, size of oil and of gas discoveries, and average wildcat depth is to take account of such differences among petroleum districts.

Other differences among petroleum districts, such as variations in the quality of the crude to be found and in its transportation and gathering costs, have been taken into account through inclusion in the analysis of the wellhead price of crude. For higher quality crude commands a price premium, and districts where gathering and transportation costs are high are districts whose crude sells at a below average price. To a rational operator, the reasons for price differentials of this type are of no concern—save insofar as they shed information on whether the differentials can be expected to continue. As everywhere in a decentralized price system, the price alone provides a sufficient signal for decision-making.[19]

Such price effects are precisely what we are attempting to measure. We are interested in the effect of changes in price on the amount and character of wildcat drilling, regardless of how such price changes come about. There is little reason to suspect that the influence of price differentials among petroleum districts is fundamentally different from that of price differentials over time, and the relatively large variance of our independent variables over petroleum districts provides an excellent set of observations for our purposes.

Nevertheless, two differences between price effects over time and price effects over districts should be noted. First, price differences among petroleum districts are persistent ones; they tend to be maintained over a rather long period of time. Thus, insofar as differences in wildcat drilling and its results among districts reflect the effects of such price differences, they reflect long-run adjustments to price effects. Such adjustments are likely to be more complete than adjustments made within a year or so to price changes over

---

[19] The situation is no different for integrated firms. A rational management should value the crude at the wellhead at its wellhead price.

time within the same district.[20] It follows that the pooling of observations from different petroleum districts means that our estimates are largely long-run ones.

This is by no means undesirable. The long-run sensitivity of wildcat operations to economic incentives is clearly of greater interest than the short-run sensitivity thereof. In fact, it is precisely the inclusion of observations over time that raises such problems as are present. The inclusion of these observations which involve short-run sensitivities is necessary, however, in view of the limited number of petroleum districts—too few to permit a purely cross-section treatment of the problem. Since adjustments of wild-catting decisions to changes in economic incentives do not require a great deal of time—not nearly so much, for example, as would be required to build or depreciate large industrial plants—the difference between long- and short-run sensitivities is not likely to be great.

On the other hand, while all this is certainly true as regards the direct effects of crude price on wildcat drilling, it is less certainly so as regards its indirect effects through its influence on other variables which in turn influence drilling. Those effects come about when the wildcat drilling induced by a rise in price leads to a deterioration of the characteristics of the average prospect drilled, because operators are willing to accept greater risks and smaller size at a higher price than at a lower one. The result of such a deterioration in past drilling results may then be to provide a disincentive to wildcat drilling in the following and later years which partly offsets the direct effects of the price increase.

It follows that the full long-run effects of price may be rather different from its relatively short-run direct effects. This will be discussed on pages 36–37, after our results have been presented. For the time being, it suffices to note that it is not particularly relevant to the present discussion of the appropriateness of pooling observations from different petroleum districts. The indirect effects just mentioned are taken into account in the analysis by letting current drilling and results depend in part on the results of last year's exploration. What is at issue here is whether the direct effects of the explanatory variables in our regressions are likely to differ greatly across districts from their magnitudes across time. As just stated, it does not seem likely that this is the case.

A further difference between adjustments over districts and over time may compensate for the understatement of direct price effects caused by short- versus long-run effects. Geographical differences in the price of crude oil represent a wider range of substitution possibilities than do temporal differences. Suppose, for example, that there are two districts which begin by being exactly alike in all respects including crude price. Suppose that for some reason the price of crude rises in one of these and that the resulting

[20] See F. M. Fisher in association with C. Kaysen, *A Study in Econometrics: The Demand for Electricity in the United States* (Amsterdam: North-Holland, 1962), pp. 126–28, for the discussion of a similar but more acute situation in another area.

price differential is maintained over time. Clearly, drilling in the high-price district will go up and drilling in the low-price district will go down. And the relative effect of the price differential across districts will be greater than the relative increase in drilling that would occur in both districts taken together if the price rise had occurred equally in both. It follows that the across-district relationship has a greater sensitivity to price (and other factors) than the long-run temporal one. Thus the pooling of observations from different districts tends to *overstate* the long-run direct sensitivity of wildcat drilling to economic incentives.[21]

While little can be done toward obtaining purely long-run estimates—the estimates of primary interest—something can be done to obtain short-run ones. The procedure is to introduce dummy variables[22] which take account explicitly of differences among petroleum districts and of nothing else, leaving other variables more or less free to represent purely short-run influences. This somewhat artificial device is used despite its inherent roughness, because it can also take account of any other differences among petroleum districts that have not already been included. Chief among these are likely to be differences in the cost of drilling and development which are not reflected by differences in depth or by differences in field size and the like. Such differences clearly do exist,[23] although it is difficult to say how important they are in the present context. Fortunately, our results seem qualitatively insensitive to such effects.

6. *Data and Notation*

The variables of our model are linked below with a discussion of the data used and with the symbols employed in presenting the results.

(1) The number of new field wildcats drilled in year t in Petroleum District j ($j = 1, \ldots, 5$) is denoted by W_{jt}. The choice of new field wildcats, rather than new field and new pool wildcats together or all exploratory wells, is a somewhat arbitrary one, as is the definition of such wells themselves. New field wildcats are chosen as the measure of exploratory effort most nearly comparable with the estimates of new discoveries used. In any case, new field wildcats make up the bulk of exploratory drilling. Data were taken from the *Bulletin of the American Association of Petroleum Geologists*.[24]

---

[21] Cf. Fisher with Kaysen, *loc. cit.*

[22] Variables which take on the value 1 from observations for a given petroleum district, and the value 0 otherwise.

[23] On differences in drilling costs, for example, see the discussion in the second essay in this volume.

[24] This source was used to ensure apparent comparability with the figures on discovery size discussed below. In computing average depth of wildcats, data from the *Oil and Gas Journal* were used. This assumes that average depth as estimated is not affected by the slight difference in coverage.

(2) The success ratio (the number of productive new field wildcats divided by the total number of new field wildcats) for Petroleum District j in year t is denoted by F_{jt}. Data are also from the *Bulletin of the American Association of Petroleum Geologists*.

(3) Average size of oil discovery per productive wildcat in Petroleum District j in year t is denoted by S_{jt}. It is in the measurement of this variable that our first really serious data problem is encountered.

Ideally, one wants both a measure of size of field as estimated by wildcatters and a measure of actual field size. The former, however, is clearly impossible to obtain and the latter raises certain problems. Even after a wildcat strikes oil, the amount of oil in the field remains unknown for several years, becoming known more and more precisely as the field is outlined and development wells are drilled.[25] For many fields, this process goes on for a considerable length of time. Indeed, the distinction between initial discoveries and later additions is often a fairly arbitrary one.[26] Moreover, the amount of oil recoverable from a given field—the really relevant variable—depends on the recovery techniques known and employed over the life of the field. Still, wildcatters do form judgments as to field size, even if with large variances, and some measure of field size must be used.

The most readily available figures on size of new discoveries can be computed from the American Petroleum Institute's figures on proved reserves. Unfortunately such a measure, however useful for other purposes, would be pointless in the present analysis. Proved reserves are known to be a conservative measure of recoverable oil, representing a working inventory of oil in the ground. They clearly understate operators' estimates of recoverable oil. Moreover, the series on additions to proved reserves through new discoveries cannot be altered to reflect a more realistic picture, since the bulk of additions to new reserves—through extensions and revisions—are not credited back to the year of field discovery but are aggregated over all known fields in the covered area. This effectively prevents any use being made of proved reserve figures in studies of discovery.[27]

---

[25] The fact that the operator of the initial discovery wildcat does not generally have full rights to the field is not relevant here so long as his returns are proportional to field size—a not unreasonable assumption. Even if this is not the case, all that is really required is that wildcatters be more interested in finding big fields than small ones, and that the size of structures can be roughly guessed in advance.

[26] See National Petroleum Council, *Proved Discoveries and Productive Capacity of Crude Oil, Natural Gas, and Natural Gas Liquids in the United States*, Report of the National Petroleum Council Committee on Proved Petroleum and Natural Gas Reserves and Availability (Washington, May 15, 1961).

[27] A number of articles by H. J. Struth (generally annual) in different oil journals do make such credits; however, the usefulness of these estimates is vitiated by the fact that Struth does not explain how the estimates are made. Aside from this, they are nationally aggregated and hence cannot be used in the present work. Struth's figures do present one feature which would be of great value if they could be used; they can be read as an annual record of credits for use in a study of the effects of development drilling and changes in recovery techniques—an impossible undertaking with any other data now in existence.

The series coming closest to that which we should like to have is one computed from the series on discoveries credited back to the year of discovery by petroleum districts, estimated by the National Petroleum Council.[28] For fields discovered before the beginning of 1955, the National Petroleum Council series gives total production plus total remaining oil recoverable by techniques known and applicable as of the end of 1959. (Fields discovered during 1955–1959 were still not well enough defined to allow such estimates to be made.) Since production techniques have changed, the series has a slight built-in bias if used as a measure of discovery size for the earlier years of our period of study. To the extent that this bias is present for the postwar years, it is largely taken into account by the inclusion of the lagged discovery size variable.

As the series on S_{jt}, then, the National Petroleum Council figures are divided by the number of productive new field wildcats. In so doing, a variable is obtained which measures average size of oil discoveries (in thousands of barrels) per new field wildcat. It is assumed that this variable is highly correlated with the average estimates of the similar figure made by wildcatters in the year after discovery, so that it can also be used to represent such estimates. To the extent that one is dealing with relationships across petroleum districts, this offers no great hazard, because estimates of discovery size are likely to be highly correlated with actual size over different areas.

(4) Average discoveries of natural gas (in millions of cubic feet) per productive wildcat are denoted by N_{jt}, the subscripts being as before. The companion series on natural gas discoveries given by the National Petroleum Council is used for the same reasons as those indicated for oil.

(5) Average depth (in feet) of new field wildcats is denoted by D_{jt}. No distinction is made between dry and productive wildcats. To do so would unduly complicate the analysis, and since the average depths of the two are highly correlated nothing would be gained from such a distinction. The figures were computed from those given in the *Oil and Gas Journal*.

(6) Geophysical and core drilling crew time is denoted by H_{jt}. The figures are computed from those on average number of crews given in *World Oil*. They are in average number of crews or crew-years. For 1946 no figures are given for Illinois and North Louisiana, and six crews in the total are unallocated. The crews have therefore been allocated between the relevant petroleum districts (II and III) in a one-to-five split, this being an approxi-

---

[28] *Op. cit.* It is hard to overemphasize how useful the National Petroleum Council study is for work of the present type and for many other purposes. One can only hope that the series will be kept up to date on an annual basis, a task of considerably less magnitude than the original undertaking. (See also p. 38, n. 52.)

mately proportional division of crews as reported for the other states in those petroleum districts.

(7) The price of crude oil at the wellhead is denoted by P_{jt}. It is computed by taking money price (in dollars) at the wellhead as reported by states in *Petroleum Facts and Figures* and the *Minerals Yearbook,* and computing a weighted average for each petroleum district, the weights being production of crude by states. This money figure is divided by three alternative price indices. The first of these is the Bureau of Labor Statistics wholesale price index (1947–49 = 1.00), and the other two are computed from the Independent Petroleum Association of America's (IPAA) drilling cost indices.[29]

The latter two deflators require a little discussion. The IPAA index is made up of an index of prices of items directly purchased by the operator, corrected for depth changes, plus an index of contractor rates (day rates are not available before 1957 and only footage rates are used). The two indices are averaged, the first receiving 54.7 per cent of the weight if 1947–49 weights are used and 69 per cent if 1959 weights are used. (The latter statement applies to the index as revised.) As depth is being explicitly included in our analysis (and in the form indicated by the second study in this volume, on drilling costs), the depth-corrected index is not used here;[30] instead, the revised index of operator costs is combined with the unrevised index of contractor rates, using alternatively both sets of weights just listed (1947–49 = 1.00). The fact that the indices start in 1947 means that such is also our starting date when these deflators are used.[31]

(8) The price of natural gas at the wellhead is denoted by G_{jt}. It was computed in a similar fashion to that just described, using the same sources. A three-year moving average is used, but the unaveraged figure was also tried without substantial change in the over-all results.

It is, of course, the case that the appropriate price variable here is the price for new contracts rather than price at the wellhead. Presumably, the gas price (if any) which is relevant for exploratory activity is that which will be paid for any gas discovery made, and this must surely be more closely related to the price under new contracts than to average value at the wellhead in all existing fields. However, new contract prices are relatively difficult to obtain for the period and geographical breakdown of this study.

---

[29] *Report of the Cost Study Committee,* various dates.

[30] The correction takes the form of *enlarging* the index when deeper wells were drilled, thus introducing the effects of depth changes rather than taking them out as required for our purposes.

[31] The revised index of contractor rates is not available before 1957. Being unable to obtain separately 1947, 1948, and 1949 values for the contractor rate index, the procedure adopted is to take for each such year the ratio of average hourly earnings in crude petroleum and natural gas production as given in the *Statistical Abstract* to the average 1947–49 figure. This produces a set of three figures which seem quite in line with the later figures of the contractor rate index. The result enters the composite index with less than half of the weight in any case and is doubtless approximately right.

Moreover, the data which are readily available for parts of seven states[32] show a very high correlation between deflated average field prices and (essentially) deflated new contract prices. It follows that the use of new contract prices instead of wellhead prices would not alter our results, since in this respect they show that natural gas prices had essentially no effect on exploratory activity during the period studied. Since this result seems entirely reasonable in view of the weak state of the market for newly discovered natural gas during the period, there seems little reason to suppose that our analysis would be changed by the substitution of new contract prices as the price variable for natural gas.

(9) Two different measures of production restrictions were tried. The first of these was the difference between scheduled allowables and calendar day allowables in Texas (averaged over months) and the second the number of Texas shutdown days (averaged over months). Availability of figures was a factor governing the sole use of Texas; another was the fact that adjustments in Texas tend to be large and to take up the greater part of whatever adjustments are thought to be necessary for the industry as a whole. The two different forms were tried because the first represents a crude measure of absolute excess capacity in crude production, while the second provides a measure of percentage excess capacity, thus providing a rough measure of the restrictions on the average operator. It should be noted, however, that production restrictions were not nearly as onerous in the 1946–1955 period as they have since become. This is of some importance in assessing our results with regard to the effects of such restrictions.

To account for possible differences in effects between Petroleum District III (which includes Texas) and other districts, each of the two variables just described was split into two parts as follows: Let A_t represent the difference between scheduled and calendar day allowables in year t; let X_t represent the number of shutdown days in that year. As before, letting the first subscript stand for the petroleum district, define:

$$A_{jt}^1 = \begin{cases} \log A_t \text{ if } j = 3 \\ 0 \text{ otherwise} \end{cases}$$

$$A_{jt}^2 = \begin{cases} 0 \text{ if } j = 3 \\ \log A_t \text{ otherwise} \end{cases} [33]$$

$$X_{jt}^1 = \begin{cases} X_t \text{ if } j = 3 \\ 0 \text{ otherwise} \end{cases}$$

$$X_{jt}^2 = \begin{cases} 0 \text{ if } j = 3 \\ X_t \text{ otherwise} . \end{cases}$$

[32] Arkansas, Louisiana, Kansas, Mississippi, New Mexico, Oklahoma, and Texas. The data were kindly supplied by J. Rhoads Foster, of Foster Associates, Inc.

[33] The reason for using logarithms here and not in the case of X_t is principally one of convenience. All logarithms are natural.

In both cases, if there is no difference between the effects on Petroleum District III and the effects on other districts, the coefficients in the regressions will be the same for A^1_{jt} and A^2_{jt} and the same for X^1_{jt} and X^2_{jt}.

(10) Finally, to allow for differences between petroleum districts not otherwise included, four dummy variables are introduced, denoted Z_1, . . . , Z_4. These are defined as follows:

$$Z_j = \left\{ \begin{array}{l} 1 \text{ for observations from the } j\text{th petroleum district} \\ 0 \text{ for all other observations} \end{array} \right.$$

$$(j = 1, \ldots, 4).$$

The effect of these dummy variables is to shift the constant term of any equation in which they appear by a different amount for each petroleum district other than the last. They thus shift the level of the dependent variable in any such equation by different amounts for each of the first four petroleum districts. When all such dummies are zero, the equation thus refers to Petroleum District V. It should be noted that if two or more of the other districts differ from Petroleum District V in the relation being estimated but do not differ from each other, it would be preferable to use one of them as the base district and thus to economize on variables. However, this will show up in the coefficients of the Z_j as estimated.

One last item remains to be discussed before proceeding to the results. In the virtual absence of information concerning the form of the functions to be fitted, natural logarithms of most variables are used for ease of interpretation.[34] This rule is not followed for $Z_1, \ldots, Z_4, A^1_{jt}, A^2_{jt}, X^1_{jt}, X^2_{jt}, H_{jt}$, or D_{jt}. In the case of $Z_1, \ldots, Z_4$ and A^1_{jt} and A^2_{jt} this is because of the way in which they are defined (the A^i_{jt} already have logs in them); X_t is not used in logarithmic form because of the presence of a zero observation, as is also the case with H_{jt}; while D_{jt} is not so used because the study of drilling costs, which is the second essay in this volume, makes it clear that the effect of depth on costs is better approximated by an exponential function than a power one.[35] With this one exception, however, there is no reason to believe that other approximations would not perform equally well with essentially the same results.

For convenience, the symbols used are summarized below.

---

[34] The coefficient of a given independent variable in a logarithmic regression equation measures the elasticity of the dependent variable with respect to that variable—the percentage change in the dependent variable per one per cent change in the independent one, other things being equal. This is a convenient measure of the sensitivity of one variable to another.

[35] All *dependent* variables are in logarithms.

SUMMARY OF NOTATION

Symbol	Explanation
j	Subscript denoting petroleum district.
t	Subscript denoting year.
W	Number of new field wildcats drilled.
F	Success ratio (ratio of productive to total new field wildcats).
S	Average size of oil discoveries per productive new field wildcat (thousands of barrels).
N	Average size of natural gas discoveries per productive new field wildcat (millions of cubic feet).
D	Average depth of new field wildcats (feet).
H	Geophysical and core drilling crew time (crew-years); a measure of information-gathering activity and scale.
P	Deflated price of crude oil at the wellhead (constant 1947–49 dollars).
G	Price of natural gas at the wellhead (three-year moving average; constant 1947–49 dollars).
A^1, A^2	Scheduled minus calendar day allowables in Texas (see pages 19–20).
X^1, X^2	Texas shutdown days (see pages 19–20).
$Z_1, \ldots, Z_4$	Dummy variables distinguishing among petroleum districts (see page 20).

7. The Principal Results and Their Interpretation

We come then to the results of this analysis. In all the regression equations presented, the numbers in parentheses are standard errors and R^2 is the square of the multiple correlation coefficient. In each case, a discussion of the effect of variables not explicitly included in the regression is presented. The effects of the dummy variables are separately discussed.

Let us begin with the equations for the number of wildcats, the success ratio, and the average size of oil discoveries per productive wildcat, and follow with a discussion of the results obtained from all three. Using the wholesale price index as a price deflator, and 1946–1955 for the equations for the first two variables and 1946–1954 for that for the third, we obtain:

(1) $\log W_{jt} = 8.29^{aaaa} + 0.00862^{aaaa}\, H_{jt} + 2.85^{aaaa} \log P_{jt}$
　　　　(1.54)　　(0.000724)　　　　(0.525)
　　　　$+\, 0.440^{aaaa} \log S_{jt-1} - 0.941^{aaa} \log F_{jt-1}$
　　　　(0.0985)　　　　　(0.304)
　　　　$-\, 0.563^{aaaa} \log N_{jt-1}$
　　　　(0.119)
　　　　　　$R^2 = .839^{aaaa}$　　　　　Degrees of freedom (d.f.) $= 44$.

　　　　　　　　a Significant at 5% level.
　　　　　　　　aa Significant at 2% level.
　　　　　　　　aaa Significant at 1% level.
　　　　　　　　aaaa Significant at $\frac{1}{10}$% level.[36]

When shutdown days and lagged depth are included,[37] the results are not substantially altered as regards the effect of variables already included.

They are:

(2) $\log W_{jt} = 8.24^{aaaa} + 0.00944^{aaaa}\, H_{jt} + 2.45^{aaaa} \log P_{jt}$
　　　　(1.91)　　(0.00127)　　　　(0.624)
　　　　$+\, 0.409^{aaaa} \log S_{jt-1} - 0.717^{a} \log F_{jt-1} - 0.535^{aaaa} \log N_{jt-1}$
　　　　(0.110)　　　　　(0.330)　　　　　(0.121)
　　　　$-\, 0.000142\, D_{jt-1} - 0.00358\, X^{1}_{jt} + 0.0298\, X^{2}_{jt}$
　　　　(0.000102)　　　(0.0436)　　　(0.0234)
　　　　　　$R^2 = .851^{aaaa}$　　　　　d.f. $= 41$.

For the success ratio, we have:

(3) $\log F_{jt} = 1.99^{aaa} + 0.581^{aaaa} \log F_{jt-1} - 0.150^{aaaa} \log S_{jt-1}$
　　　　(0.651)　(0.116)　　　　　(0.0407)
　　　　$+\, 0.0830 \log N_{jt-1} - 0.000106^{aaa}\, D_{jt-1}$
　　　　(0.0448)　　　　　(0.0000375)
　　　　$+\, 0.000590^{a}\, H_{jt} - 0.356 \log P_{jt}$
　　　　(0.000283)　　　(0.234)
　　　　　　$R^2 = .729^{aaaa}$　　　　　d.f. $= 43$.

For average oil discoveries per productive wildcat:

(4) $\log S_{jt} = 5.35^{aaa} + 0.777^{aaaa} \log S_{jt-1} + 0.692^{a} \log F_{jt-1}$
　　　　(1.73)　(0.113)　　　　　(0.329)
　　　　$-\, 0.489^{aaaa} \log N_{jt-1} - 2.18^{aaa} \log P_{jt}$
　　　　(0.137)　　　　(0.631)
　　　　　　$R^2 = .848^{aaaa}$　　　　　d.f. $= 40$.

[36] This notation will be used for all later regression equations. The term "significance" is, of course, used in a purely statistical sense as a measure of the likelihood that the parameter being measured is actually zero rather than as equivalent to "important" or "large."

[37] The results are not substantially different when these are included separately. The shutdown-days variables seem to work slightly better than does the difference between scheduled and calendar day allowables.

In neither of the last two equations does the introduction of further variables appear to yield anything of interest, save as discussed below.

What, then, can we make of these results? Clearly, most of our coefficients are significant, but more important, the coefficients themselves indicate a reasonably clear and consistent behavioral pattern.[38]

In the first place, there is clear evidence of the effects of price. The elasticity[39] of wildcat drilling with respect to the price of crude is about $+2.85$ from equation (1) and $+2.45$ from equation (2), indicating a sizable price effect. Furthermore, the argument above that price should also affect the characteristics of accepted prospects is strongly supported. The elasticity of the success ratio with respect to price is about -0.36 and, while the coefficient in question is not significant, it has the expected sign (indicating a worsening of prospect characteristics when prices rise). Moreover, the really large effect here appears to be on discovery size which has a price elasticity of about -2.2, the coefficient being significant at the one-tenth of one per cent level.

These results are not unreasonable. It seems plausible to suppose that the underlying size distribution of prospects is highly skewed toward small ones so that a small price change greatly changes the number of *small* prospects which are deemed worth drilling; on the other hand, the distribution of risk over prospects seems less likely to be so skewed, for risk tends to be reduced by information-gathering activity before drilling is seriously considered. Further, operators may prefer a diminution of size to an increase of risk when price rises. Finally, as explained on pages 25–27, there tends to accumulate a set of undrilled prospects about which something is known, which set consists principally of relatively small, relatively certain prospects. It follows that a rise in price induces a decrease in average size associated with an *increase* in certainty which partially offsets the risk increase which would otherwise occur. The effect is a short-run one and is restricted to price rises.

The clear implication of the above is that the sensitivity of new oil discoveries to economic incentives is substantially less than the similar sensitivity of wildcat drilling. This is primarily caused by the deterioration of discovery size which comes about when small prospects are made attractive by a price increase.

The effects of geophysical and core drilling crew time also seem plausible. As expected, in the regressions for W_{jt}, where H_{jt} plays the role of a scale variable, the coefficient thereof is significant and positive and the elasticity involved is somewhat less than unity.[40] There is also some effect (although it

---

[38] This is far more important than mere formal significance. See F. M. Fisher, *A Priori Information and Time Series Analysis*, p. 16; and F. M. Fisher in association with C. Kaysen, *A Study in Econometrics: the Demand for Electricity in the United States*, Chapter 1.

[39] See footnote 34.

[40] Recall that H_{jt} is not in logarithms so that its coefficient is not unit free as are most of the others presented.

is quite a small one—elasticity being very roughly about +.2) of geophysical crew time on the success ratio. This doubtless reflects the acquisition of information about prospects. No appreciable effect on discovery size is apparent.

The effects of depth are also easy to interpret. Depth tends to be highly auto-correlated, deep wildcats last year usually being associated with deep ones this year. This is because with greater depth the large rate of increase in cost tends to promote a thorough working of relatively shallow prospects before relatively deep ones, other things being equal. It is therefore no surprise to find a slight negative effect on wildcat drilling (although the coefficient is insignificant). Similarly, since uncertainty increases with increasing depth, it is natural to observe a negative (and significant) effect of depth on the success ratio when information-gathering activity is held constant.

The effect of the shutdown days variable, on the other hand, seems to vary in different petroleum districts (although both coefficients are insignificant). The principal effect appears to be a shifting of wildcatting from Petroleum District III to other areas when shutdown days rise. However, this result must be accepted with caution. Production restrictions during the period being examined were not nearly so severe or so prolonged as they have recently been. It may well be that there is little over-all effect on wildcatting from such restrictions, so long as they are fairly light, but a sizable effect when they are heavy. This is not only because the restrictions themselves may be onerous (in part this is picked up by the shift from Petroleum District III), but because a high number of shutdown days is an indication of serious overcapacity in the industry at current prices. Without more recent data on some of the variables, it is impossible to extend the analysis to find out about this. (There is somewhat stronger evidence of an effect of restriction of allowables on geophysical crew time—that is, on the number of prospects considered—and thus indirectly on wildcat drilling. This will be discussed in Section 8.)

So far, this has been reasonably straightforward. We now come, however, to a far trickier problem, that of interpreting the coefficients of the three principal characteristics of past prospects accepted, S_{jt-1}, F_{jt-1}, and N_{jt-1}. This is a bit more complicated because these variables do triple duty in our analysis. They distinguish between petroleum districts, they are involved in the underlying process which determines prospect characteristics, and, as indications of what wildcatters can expect, they have economic influences. The hypotheses which will be put forward were in part suggested by the results obtained. They do, however, seem otherwise plausible and do explain not one coefficient but several. Moreover, it happens that in the course of the study different regressions were run at different times. The hypotheses presented were suggested by results at an early stage, and the fact that their implications as to the coefficients to be expected in later regressions were borne out is thus a strong favorable point. Nevertheless, they are obviously fit subjects for further study and testing.

Size versus Certainty and the "Inventory" of Small Prospects

The effects of S_{jt-1} are fairly clear. Large discoveries tend to act as an incentive to wildcatting, other things being equal, so that there is a significantly positive coefficient of log S_{jt-1} in equations (1) and (2). This statement should perhaps be interpreted as applying to a long-run effect, i.e., districts with large past discoveries tend to have more wildcatting than districts with small ones, *ceteris paribus*. Similarly, the significantly positive coefficient of log S_{jt-1} in equation (4) is easy to interpret. First, that term in that regression is a scale variable; it distinguishes characteristics of petroleum districts with regard to average size of oil discoveries. Thus a positive coefficient is in part due to the fact that districts which had large discoveries in the past are likely to have large ones in the present. However, there is an additional effect.

The large prizes apparently offered by large discoveries in such districts and in areas within them may promote a tendency to drill relatively more for size and less for certainty there than in districts and subareas where fields historically have been relatively small. The result of this is to add something to the positive coefficient of log S_{jt-1} in equation (4). More important than this result, however, is the implication for the corresponding coefficient in equation (3). Large prospects tend to be relatively uncertain ones, either because certain types of small geologic structures are more likely to contain oil than are big structures, or because large prospects, being more attractive, tend to be drilled when relatively less is known about them than about small ones. The result is that any incentive which enhances the attractiveness of large fields as such tends not only to increase discovery size per *productive* wildcat but also to decrease the success ratio. Hence we obtain the small but significantly negative coefficient of log S_{jt-1} in the equation for the success ratio (3).

We turn, then, to the effects of the past value of the success ratio, F_{jt-1}. Here the immediate impression is likely to be one of difficulty. Is it not odd that we obtain a significantly negative coefficient in equations (1) and (2)? Surely an increase in the ease of finding oil should increase wildcatting, not decrease it. The difficulty is only apparent, however, and stems from confusion between the probability of finding oil as viewed *ex ante* and the success ratio as viewed *ex post*. As has been pointed out on pages 6–7, a rise in the success ratio last year does not necessarily mean a rise in the ease of finding oil on prospects considered either last year or this year; rather, it means an acceptance of relatively certain prospects last year, and this is by no means the same thing.

It is clear that, other things being equal, operators will drill a large prospect rather than a small one, at the same degree of risk. On the other hand, the passage of time and the collection of information can reduce the degree of risk associated with a prospect, but cannot change its size. It follows that at any given price level, if we consider the collection of prospects about which something is known but which are too small or too risky to be worth drilling, that collection will consist largely of relatively small, relatively certain prospects, given any reasonable assumption about the distribution of prospects in nature. This is so because large prospects tend to pass out of the collection at a higher degree of risk than do small ones. It follows, first, that the distribution of available prospects tends to show a negative correlation between size and certainty and, second, that it contains a possibly large tail of relatively small, relatively certain prospects in part as a sort of inventory from earlier years,[41] and in part as a reflection of the probable skewness of the underlying size distribution of oil fields already discussed.

This has a number of consequences, the first of which we have already observed. When the price of crude rises—and the period in question was one of rising prices—the availability of a large number of relatively small, relatively certain prospects means that there will be a relatively large decline in average discovery size and a relatively small one in the success ratio. This contributes to the relative magnitude of the price coefficients in equations (3) and (4). On the other hand, the fact that relatively small and certain prospects tend to accumulate means that a year in which the success ratio is higher than usual is a year in which the accumulated inventory of such prospects is being depleted at a faster than usual rate.[42] (Note that H_{jt}, the rate of acquisition of new information, is held constant in the regressions.) The result in the following year (other things being equal) is a distribution of prospects in which the number whose expected size and risk barely justify their drilling is smaller than usual. Consequently, the number of prospects drilled is smaller than it would otherwise have been. Hence the significantly negative coefficient of log F_{jt-1} in equations (1) and (2). Further, if the inventory of prospects does indeed have a large number of relatively small prospects and if the size distribution of such prospects is rather flat in the neighborhood of just acceptable size, then a small change in F_{jt-1} will be associated with a rather large change in the inventory. There

---

[41] The reference here to a "prospect inventory" does not mean that any one firm need repeatedly consider the prospects in question at the final decision level. The prospects are those about which a fair amount may be known but which have not been thought worth passing up to top management for decision at the given price level. On the existence of such a prospect inventory, see Grayson, *op. cit.*, pp. 42–43. The author is indebted to Robert B. Wilson for this reference.

[42] This would not be the case if the year in question were one in which a large new undrilled and previously unavailable area was opened up. This did not occur during the period studied, offshore fields becoming available at the end of that period. The author is indebted to C. T. Leenders for the general point.

is thus nothing disturbing about the absolute magnitude of the coefficients of F_{jt-1} in equations (1) and (2).[43]

In addition, this hypothesis accounts for the significantly *positive* coefficient of log F_{jt-1} in the regression for average discovery size (4).[44] The effect of inventory depletion is to reduce the number of *small* prospects that would otherwise be drilled. This means that the average size of discoveries per productive wildcat rises.

Finally, all this implies that a rise in F_{jt-1} is accompanied by a fall in F_{jt}, other things being equal. We cannot expect to observe this effect directly, however, because the lagged success ratio serves as the primary scale variable distinguishing among petroleum districts in the equation for the current success ratio (3). Districts in which (other things being equal) oil and gas have been relatively easy to find remain such districts. However, there is some mild further evidence for the effect in question in the fact that the coefficient of log F_{jt-1} in equation (3), which includes both the inventory-depletion effect (negative) and the district-distinguishing effect (positive), is somewhat smaller than the coefficient of log S_{jt-1} in equation (4), which includes both the district-distinguishing effect (positive) and the incentive-toward-larger-prospects effect (positive) already discussed. This is only a mild piece of evidence because there is little reason to suppose that the district-distinguishing effect in the two cases is of the same magnitude. Nevertheless, the observed result is clearly consistent with the hypothesis.

The Role of Natural Gas

Some further favorable evidence appears in the results for natural gas discovery size. Clearly, the primary items in these results which require explanation are the significantly negative effect of past gas discovery size on wildcatting (equations [1] and [2]) and on average oil discovery size (equation [4]), as well as the absence of the price of gas as an important variable.

Historically, the discovery of natural gas in the United States was a byproduct of the search for crude petroleum, and it is only relatively recently that natural gas has become a valuable discovery in its own right. Moreover, the increase in the importance of natural gas was only partly reflected in relative price trends. The price of gas only reflects the value of gas when sold, while a considerable part of the problem with natural gas was the inability to sell non-associated gas at all for a number of years after its discovery for want of a market and of transportation facilities. This situa-

[43] The author is indebted to Anders Ølgaard for pointing this out.

[44] Note that this coefficient is positive, despite the argument as to the negative correlation between size and certainty and despite the negative coefficient of S_{jt-1} in equation (3). There is a *real* effect to be explained here and the hypothesis under discussion does so.

tion has implications for our results which clearly cannot be extended to the present time.

Consider two regions with the same average probability of finding either oil *or* gas and the same level of lagged average size of oil discoveries per productive wildcat. Suppose that discoveries of natural gas per productive wildcat have been considerably larger in one region than in the other. It is not unreasonable to suppose that the region with the higher historical gas-to-oil ratio has a slightly lower expected value of future size of oil discoveries per productive wildcat than the region with the lower historical gas-to-oil ratio. This may be so because it is true of the underlying distribution of such resources, or because there is simply a greater probability of finding further gas fields in the region that has produced more gas and thus adding nothing to total oil discoveries while adding a productive wildcat to the denominator of S_{jt}. Since the probability of finding either oil or gas are held constant—so that the operators' estimates of the probability of finding one or the other are the same—we should expect to see a (small) negative coefficient of log N_{jt-1} in the regression for S_{jt} (4).

Furthermore, since gas was substantially less valuable than oil in the period studied, other things being equal, the attractiveness of drilling in a high-gas region was likely to be less because the probability of finding not oil but gas was higher. Moreover, a region with relatively high gas discoveries in the past was likely to be one in which the time period required to market further gas discoveries is relatively long. Both these factors imply that (other things being equal) there was an economic effect of past gas discoveries which reduced the incentives to drill. This makes the coefficient of log N_{jt-1} in equations (1) and (2) negative.

Moreover, if past gas discoveries have been of large size, this may be taken by operators as an indication that *large* structures in the given district are relatively likely to contain gas. This could happen if, for example, the areas within the district which contain large structures also have a high percentage of gas. The result may then be to reduce the attractiveness of drilling on such structures or in such areas and hence to reduce the average size of oil discoveries per productive wildcat. This effect reinforces those described above and also acts to make the coefficient of log N_{jt-1} in the regression for S_{jt} (4) negative.

Finally, since large prospects and certainty tend to be negatively correlated, a side effect of this should then be an increase (other things being equal) in the success ratio. This accounts for the positive coefficient of log N_{jt-1} in equation (3), although that coefficient is not quite significant—which is not surprising because the effect is rather indirect.

If this argument is correct, it has certain further implications which appear to be borne out; some of them also bear on the earlier hypothesis as to the inventory-depletion effect associated with the lagged success ratio.

In the first place, one might expect the economic effects just described to be less the higher the ratio of gas-to-oil prices, although this may not be the case in view of the difficulty of marketing gas. To test this, suppose that the true equation in each case included not simply a term in the logarithm of N_{jt-1} but a term of the form

$$\left(\alpha + \beta \, \frac{G_{jt}}{P_{jt}} \right) \log N_{jt-1} \, .$$

The hypothesis would then imply a negative α and a positive β in the regressions for W_{jt} and S_{jt} and the reverse in the regression for F_{jt}. This is indeed observed in the first two regressions when such a term replaces that in log N_{jt-1} alone, but β is *very* small in absolute value and is insignificant. In the success ratio regression β is very slightly positive and is insignificant. In all cases, the estimated β's were much smaller than their standard errors, so that the evidence here is mildly favorable but extremely slight. Since the price at which gas is being sold is irrelevant for an operator whose market for gas lies several years in the future, the null effect of current gas price is unsurprising.[45]

Verifying Some Predictions

Much stronger favorable evidence is provided for our hypothesis by the regression for the current average size of gas discoveries per productive wildcat.[46] It was conjectured above that, given the past value of S_{jt}, and holding constant the principal factors affecting *total* discovery probability, areas with high past gas discoveries per productive wildcat might reasonably be expected to have a lower current value of oil discoveries per productive wildcat than areas with low past gas discoveries. Similar reasoning leads to the conjecture that, given the past value of N_{jt}, and holding the same factors constant, areas with high past oil discoveries per productive wildcat might reasonably be expected to have a lower current value of gas discoveries per productive wildcat than areas with low past oil discoveries. On the other hand, it has already been argued that a high past value of S_{jt} is itself an added incentive to the drilling of larger structures, other things being equal. It follows that the effect of S_{jt-1} on N_{jt} should be made up of two offsetting influences.

Similarly, it has been argued that a high past value of N_{jt} serves to reduce the attractiveness of large structures, other things being equal. It follows

---

[45] The results would be unchanged if new contract prices were used instead of wellhead prices. See the discussion on pp. 18–19.

[46] The hypothesis in question was formed before running this particular regression which was done after those already reported. The foregoing paragraphs thus represent predictions rather than forming part of an ad hoc explanation.

that the effect of N_{jt-1} on N_{jt} should also be made up of two offsetting influences, the one just described and the district-distinguishing effect.

However, if the discovery of natural gas in the period in question was principally a byproduct of the search for oil, all other effects on S_{jt} should be essentially duplicated in the regression for N_{jt}. Thus the elasticities of N_{jt} with respect to price and with respect to the lagged success ratio—the inventory-depletion effect—should be about the same as the corresponding elasticities of S_{jt} (the appropriate coefficients in equation [4]). The verification of these predictions therefore provides considerable evidence for several earlier arguments.

The predictions are in fact sharply verified by the regression, which applies to the period 1946–1954:

$$(5) \qquad \log N_{jt} = 7.93^{aaa} + 0.0368 \log N_{jt-1} - 0.0473 \log S_{jt-1}$$
$$\qquad\qquad (1.87) \qquad (0.149) \qquad\qquad (0.123)$$

$$\qquad\qquad + 0.840^{a} \log F_{jt-1} - 2.01^{aa} \log P_{jt}$$
$$\qquad\qquad (0.357) \qquad\qquad (0.684)$$

$$\qquad R^2 = .316^{aaa} \qquad\qquad \text{d.f.} = 40 .$$

The closeness of the last two coefficients to the corresponding ones in equation (4) is especially gratifying. Finally, the fact that R^2 (although still highly significant) is somewhat lower here than in our earlier results is not surprising, since, in some sense, the relation being investigated is a byproduct of another and there are forces working in opposing directions.

Alternative Results

It is further reassuring to note that there are no substantial qualitative changes in our results and that all hypotheses are supported by the same evidence as before when the wholesale price index is replaced as price deflator by the two indices constructed from the IPAA drilling cost indices. As the results are almost precisely the same no matter which set of weights is used in the construction of the index, it is necessary to present only the regressions in which the 1947–49 weights were employed. These should be compared with equations (1) to (5), respectively. The period is 1947–1955 in the regressions for W_{jt} and F_{jt} and 1947–1954 in the other two regressions.

$$(6) \quad \log W_{jt} = 8.71^{aaaa} + 0.00903^{aaaa} H_{jt} + 2.84^{aaaa} \log P_{jt}$$
$$\qquad\qquad (1.62) \qquad (0.000771) \qquad\quad (0.634)$$

$$\qquad\qquad + 0.334^{aaaa} \log S_{jt-1} - 1.27^{aaaa} \log F_{jt-1} - 0.540^{aaaa} \log N_{jt-1}$$
$$\qquad\qquad (0.0923) \qquad\qquad (0.343) \qquad\qquad (0.132)$$

$$\qquad R^2 = .835^{aaaa} \qquad\qquad \text{d.f.} = 39 .$$

(7) $\log W_{jt} = 8.03^{[aaaa]} + 0.00976^{[aaaa]} H_{jt} + 2.27^{[aaa]} \log P_{jt}$
$\qquad\quad (1.77) \qquad (0.00106) \qquad\quad (0.651)$

$\qquad + 0.278^{[aaa]} \log S_{jt-1} - 0.679^{[a]} \log F_{jt-1} - 0.375^{[aaa]} \log N_{jt-1}$
$\qquad\quad (0.0911) \qquad\qquad (0.296) \qquad\qquad (0.110)$

$\qquad - 0.000444^{[aaaa]} D_{jt-1} + 0.0297\ X^{1}_{jt} + 0.0655^{[aaa]}\ X^{2}_{jt}$
$\qquad\quad (0.000118) \qquad\qquad (0.0351) \qquad (0.0205)$

$\qquad\quad R^2 = .905^{[aaaa]} \qquad\qquad \text{d.f.} = 36 \ .$

(8) $\log F_{jt} = 1.97^{[aaa]} + 0.642^{[aaaa]} \log F_{jt-1} - 0.134^{[aaa]} \log S_{jt-1}$
$\qquad\quad (0.685) \quad (0.141) \qquad\qquad (0.0404)$

$\qquad + 0.0597 \log N_{jt-1} - 0.0000835\ D_{jt-1}$
$\qquad\quad (0.0512) \qquad\qquad (0.0000555)$

$\qquad + 0.000549\ H_{jt} - 0.393 \log P_{jt}$
$\qquad\quad (0.000294) \qquad (0.289)$

$\qquad\quad R^2 = .715^{[aaaa]} \qquad\qquad \text{d.f.} = 38 \ .$

(9) $\log S_{jt} = 4.36^{[a]} + 0.852^{[aaaa]} \log S_{jt-1} + 0.645 \log F_{jt-1}$
$\qquad\quad (1.78) \quad (0.0997) \qquad\qquad (0.337)$

$\qquad - 0.405^{[aaa]} \log N_{jt-1} - 1.63^{[a]} \log P_{jt}$
$\qquad\quad (0.144) \qquad\qquad (0.714)$

$\qquad\quad R^2 = .861^{[aaaa]} \qquad\qquad \text{d.f.} = 35 \ .$

(10) $\log N_{jt} = 7.45^{[aaa]} + 0.0735 \log N_{jt-1} + 0.0169 \log S_{jt-1}$
$\qquad\quad (2.10) \qquad (0.169) \qquad\qquad (0.117)$

$\qquad + 0.803 \log F_{jt-1} - 1.55 \log P_{jt-1}$
$\qquad\quad (0.397) \qquad\qquad (0.842)$

$\qquad\quad R^2 = .233^{[a]} \qquad\qquad \text{d.f.} = 35 \ .$

There are only a few differences from earlier results which seem worth remarking. First, in equation (7) the negative effect of depth on wildcatting appears larger than in equation (2)—about 4.4 per cent per hundred feet. The coefficient in question is also significant. In some sense, this change appears reasonable since an explicit index of drilling costs is being used.

In the same equation, the effect of shutdown days on wildcatting outside Petroleum District III is now significant, but the effect on drilling in that district has changed sign. As it is still much smaller than its standard error, however, this requires no particular comment beyond that already given above.

The only other change which requires comment is the reduction of the estimate of the price effect on oil and gas discovery size (equations [9] and

[10] as against equations [4] and [5]) from an elasticity of about -2.1 to one of about -1.6. The general conclusion that such an effect exists and is of some magnitude remains unchanged, however. Moreover, the fact that the estimated effect is the same in equation (9) as in (10) still provides verification of the earlier discussion.

In general, the fit of the equations just presented is poorer than that of those previously given, although the difference is usually slight. The fit is better in equation (7) than in (2).

Finally, it is worth emphasizing that whichever set of equations is regarded as the superior one, the evidence for the previous discussion, and for that given below, is essentially the same.[47]

The hypotheses also seem to stand up well when the dummy variables $Z_1, \ldots, Z_4$ are introduced. Those variables, it will be recalled, have two effects. First, they distinguish among petroleum districts in ways (if any) not already accounted for; second, they make the results strictly applicable only to short-run effects. Since such effects are of small interest for our purposes, and since the equations already presented may be taken to contain the most valid results, no specific presentation of the equations with dummy variables will be made, but they will be discussed in a qualitative way. Let us begin by considering what the effects of these variables should be if previous results and discussion hold.

In the first place, for each relationship, if the within-petroleum-district means of the dependent variable are in roughly a linear relation with the similar means of one or more of the petroleum-district-distinguishing variables already included (S_{jt-1}, N_{jt-1}, F_{jt-1}, H_{jt}, and D_{jt-1}), the dummy variables will essentially duplicate the effect of such variables in distinguishing among petroleum districts, leaving only the other effects of those variables to be explained by them. Further, since the resulting relationship will be short-run, any such explanatory variable whose effect is principally long-run will disappear from the equation.

This is essentially what is observed. The dummy variables in all cases appear to act as substitutes for past oil and gas discovery size in distinguishing among petroleum districts, and the coefficients of the latter variables become negligible. (Since the economic effects of those variables discussed above seem likely to be long-run, this is reasonable.) On the other hand, it cannot be said that there is convincing evidence for the presence of other district-distinguishing effects than those explicitly accounted for, as there is no case in which all four dummy variables are significant at the same time —and sometimes none are—the identity of the significant ones varying from equation to equation. The indications are thus that the dummy variables substitute for the discovery size variables in distinguishing among petroleum districts, but add little else in this regard.

---

[47] Although it is perhaps a bit stronger in the case of equations (1) to (5) in view of the significance levels of various coefficients.

The results as regards economic effects are also as expected. It seems reasonable that the effects of past oil and gas discovery size on incentives are relatively long-run ones and hence should disappear. On the other hand, the argument on pages 25–27 as to the inventory-depletion effect of the success ratio was clearly an argument as to a short-run effect, and it is thus further evidence in favor of that argument that the coefficients which support it are practically unchanged in the regressions under discussion. Finally, as was argued on pages 13–15, direct price effects are a mixture of short- and long-run influences and, as this implies, the introduction of the dummy variables reduces the magnitude of all price coefficients somewhat without changing their sign. Thus the elasticity of wildcat drilling with respect to price falls to about +1.5 and that of oil discoveries per productive wildcat to about −.6. The price elasticity of the success ratio is reduced only negligibly, while that of gas discoveries per productive wildcat is reduced to about −1. In all cases, standard errors are increased although the direct price effect on wildcatting remains highly significant. This is reasonable, since there is clearly a great deal more randomness in any short-run direct effect of price on discoveries than in the corresponding long-run effect, while this is clearly less true of the effect on wildcat drilling itself.

In all cases, there is nothing to indicate that our arguments or conclusions should be changed as a consequence of this experiment. Rather, they are strengthened.

8. The Sensitivity of Wildcat Drilling and of New Oil and Gas Discoveries to Economic Incentives

We now return to the main problem with which this study is concerned, that of the sentivity of wildcat drilling and of new oil and gas discoveries to economic incentives. Before discussing the interpretation of our results in this context, an important negative result must be examined, since it bears directly on the problem being investigated.

As is evident from either equations (1) and (2) or (6) and (7), the use of geophysical crew time as a measure of the scale of activity—of the number of prospects seriously considered—has been highly successful. That variable clearly has a positive role in wildcat drilling. It is therefore possible that, aside from the effects already directly estimated, price and other economic variables also influence wildcat drilling and its results indirectly, by affecting the decision to consider certain prospects as well as the decision to drill them, once considered. In other words, geophysical crew time may itself be influenced by economic variables, this relationship being a measure of the influence of such variables on the scale of activity.

On the other hand, the number of prospects seriously considered (as measured by geophysical crew time) may equally well not be substantially affected by economic variables—at least as long as the latter variables do not move out of certain ranges. The information-gathering decision involved here is typically made earlier—and possibly by different decision makers— than the decision on which wildcat wells to drill. Indeed, the information gathered is a prime factor in the latter decision. Further, while the decision to investigate a prospect by geophysical means involves no little cost, far less expense is incurred than in the drilling of an actual wildcat. It follows that, depending on the information otherwise available, when price falls or when for other economic reasons there is a rise in the minimum level at which prospects are attractive for drilling, the decision to investigate by geophysical means may be less costly than foregoing this step and incurring losses due to lack of information.[48] It is even possible, therefore, that a rise in geophysical crew time might accompany slight declines in economic incentives, although this is clearly unlikely when any substantial deterioration of economic conditions occurs in the industry.

It is not surprising, therefore, to find only a little evidence of economic effects in this area during the period in question. The process which determines H_{jt} appears to be largely dependent on the opportunities found in nature and relatively little influenced by crude price. On the other hand, there is some evidence of an effect of production restrictions here. When H_{jt} is regressed on (a) its own lagged value (to take account of trends in the natural process involved), (b) the petroleum-district-distinguishing dummy variables (the results are more or less similar when these are omitted), and (c) the two variables for shutdown days, the results are as shown in equation (11). The period is 1947–1955 because information on crew time by petroleum districts is lacking for 1946.

$$(11) \qquad H_{jt} = 13.6 + 0.638^{aaaa} H_{jt-1} - 7.41\, Z_1 + 23.0\, Z_2$$
$$\phantom{(11) \qquad H_{jt} =} (12.2) \quad (0.141) \qquad\quad (12.4) \qquad (15.0)$$

$$+ 150^{aaa}\, Z_3 + 32.4^{a}\, Z_4 - 0.871\, X^1_{jt} - 0.508\, X^2_{jt}$$
$$(49.3) \qquad (15.8) \qquad (2.147) \qquad (1.02)$$

$$R^2 = \ .973^{aaaa} \qquad\qquad \text{d.f.} = 37 \ .$$

While the coefficients of the shutdown days variables are obviously far from being significant, their values are not unreasonable. Thus the effect in Petroleum District III appears to be greater than outside that district, a result which is compatible with our earlier findings.[49] This is the only evidence found of an economic effect; in all other cases, the coefficients involved were not significant and the effect itself was minuscule. It offers one

---

[48] See Grayson, *op. cit.*, Chapter 11, for a discussion of the decision here involved.

[49] Note that the implication of the coefficient of H_{jt-1} in equation (11) is that the long-run effect of production restrictions may be substantially greater than their short-run effect (by a factor of about three).

explanation for the decline in geophysical crew time, and hence in wildcat drilling, that has taken place in recent years coincidental with a large rise in shutdown days. Even this latter circumstance, however, may be more simply due to the properties of the underlying stochastic process generating opportunities, or it may be due to both effects.

If production restrictions do affect H_{jt} as in equation (11), there is some possibility that the sensitivity of wildcat drilling to price changes may be somewhat less when those price changes are brought about by restricting production than otherwise. This opens, but does not settle, the question of the effectiveness of policies designed to stimulate discoveries through maintaining a high price by means of production restrictions.

Returning then to the results of the last section, let us consider what light they throw on the main question. We have already observed that there are clearly direct effects of price, both on the number of wildcats drilled and on the characteristics of the average prospects accepted. To repeat, the elasticity of wildcat drilling with respect to price is estimated at +2.85 from equation (1), +2.45 from (2), +2.84 from (6), and +2.27 from (7). The elasticity of the success ratio with respect to price is −0.36 from equation (3) and −0.39 from (8); that of the average size of oil discoveries per productive wildcat is −2.18 from equation (4) and −1.63 from (9); finally, that of the average size of gas discoveries per productive wildcat is −2.01 from equation (5) and −1.55 from (10).

Now, since total discoveries of new oil are the product of wildcats drilled, the success ratio, and the average size of oil discoveries per productive wildcat, the elasticity of total new oil discoveries with respect to price is the sum of the price elasticities of those three variables. A similar statement holds for the price elasticity of total new gas discoveries. As estimated, this places the price elasticities of both new oil and new gas discoveries at a maximum of about +0.9, using equations (6), (8), (9), and (10), and considerably lower figures of about +0.3 seem quite likely since (1) to (5) may be regarded as the best set of equations. In any case, the elasticity involved is clearly far less than the elasticity of roughly +2.2 to +2.8 of wildcat drilling itself with respect to price.

This point deserves special emphasis even beyond that already given. Reading the estimates as given, the implication is that whereas a 10 per cent fall in the price of crude (or a cut in the depletion allowance with a similar effect on profits) would, in the long run, reduce wildcat drilling by perhaps a bit more than 25 per cent, the resulting effect on new oil and gas discoveries would be a reduction of only about 9 per cent at most—and quite possibly by only a little more than 3 per cent—principally because the reduction in wildcats would be largely at the expense of small prospects.

But should these results be read as given? Moreover, to what extent are they directly relevant to policy evaluation? There are a number of questions to be discussed here.

In the first place, it will be recalled (pages 13–14) that our procedure inevitably combines short- and long-run influences. While it seems clear that those that are long-run predominate in our estimates of direct price effects, nevertheless the short-run influences must have some effect. The usual expectation in econometrics is that estimated short-run direct price effects tend to be lower than long-run ones as they tend to reflect partial completion of an adjustment process. To the extent that this is true in the present case, it would primarily apply to the direct price elasticity of wildcat drilling and our results would understate—but only slightly—the true direct elasticity involved.

On the other hand, we have seen (pages 14–15) that the inclusion of observations from different petroleum districts tends toward overstatement of the direct price elasticity of wildcat drilling since locational effects additional to those which occur over time are thereby introduced. This effect is probably more serious than the short- versus long-run one just discussed.

But there is another relation in our case between short-run and long-run price elasticities which is more complicated than the rather general one just mentioned. We have found that the principal effect of a price change on the characteristics of the prospects drilled is on average discovery size rather than on the success ratio. To the extent that this is a consequence of the skewness of the underlying size distribution of fields, it is a long-run phenomenon. To the extent, however, that it is related to the accumulation of an inventory of relatively small, relatively certain undrilled prospects (as discussed on pages 26–27), it is in part a short-run phenomenon, since that inventory will be exhausted in the short run. Moreover, it is a phenomenon that would not occur in the case of a fall in price. It follows that our results somewhat *overstate* the absolute value of the long-run price elasticity of average discovery size, yielding an estimate of too large an effect.

On the other hand, the discussion as to the inventory of undrilled prospects also implies that our results somewhat *understate* the absolute value of the long-run price elasticity of the success ratio, yielding an estimate of too small an effect. Further, since the existence of that inventory adds in the short-run to the number of prospects attractive in the event of a price rise, an upward bias is introduced in our estimates of the long-run price elasticity of wildcat drilling.

Moreover, there are grounds for believing that the full direct *and indirect* effect of price on wildcat drilling is overstated by the previous discussion, at least for present relationships, for an additional reason. The price elasticity of wildcat drilling was estimated in the equations by a procedure *which held constant the lagged success ratio, and the lagged average size of oil and gas discoveries per productive wildcat.* As we have seen, however, the current values of those variables are themselves affected by price. It follows that the long-run effect of a price rise on wildcat drilling includes

both the effects heretofore discussed and also effects through the changes
in the characteristics of accepted prospects which occur as the prospect
inventory is depleted at a higher rate following a price rise. Thus, from
the equations, when price rises by 10 per cent, there is a direct effect on
wildcat drilling of very roughly 25 per cent. However, there is also a 4 per
cent reduction in the success ratio and a 20 per cent reduction in average
discovery sizes per successful wildcat. The incentives to further wildcatting
are therefore reduced below the level indicated by the direct price effect
alone, for the offsetting effect of the fall in average gas discoveries indicated
by the negative coefficients of log N_{jt-1} in the equations for W_{jt} and the
earlier discussion (pages 27–28) cannot be expected to be present any longer
in view of the great increase in the importance of natural gas since the end
of the period for which these equations were estimated.

It is impossible, however, to assess the magnitude of such indirect effects
precisely from the equations. It is not only the changing role of gas
that is involved; but also the reduction of wildcat drilling due to such indirect effects of a price rise would in turn serve partially to raise the success
ratio and average discovery sizes, because the cutback would—as in the general case repeatedly described—be applied to worse prospects than would
otherwise be drilled. This in turn would increase the incentive to drill, other
things being equal. The equilibrium of this process lies somewhere between
that indicated by the direct effects and zero.[50]

It follows, then, that our results overstate the full long-run effect of price
on wildcat drilling. In the long run, that effect will be lower than would
be indicated by a literal reading of the numerical results. In a period of
rising prices such as that from which the data for this study were taken, the
result of the inventory depletion effect is a greater price sensitivity of wildcat drilling in the short run than in the long run. On the other hand, as
already remarked, the presence of that effect also leads to an overstatement
of the long-run sensitivity of discovery size to price and an understatement
of the long-run price sensitivity of the success ratio. Since the presence of
an inventory of undrilled prospects provides a readily accessible collection
of marginally attractive prospects which are of greater relative importance
in the short than in the long run, there seems little doubt that the numerical results overstate the difference between the long-run price elasticity of
wildcat drilling and the long-run price elasticity of new discoveries. That
such a difference persists even in the long run, however, is clear since an
increase in price inevitably makes attractive prospects which would not
otherwise be so.

It is not clear, however, that the sensitivity of wildcat drilling and of new
oil (and gas) discoveries, as measured by our results, are immediately rele-

---

[50] We do not solve the estimated equation system to determine such an equilibrium numeri-
cally, as the dynamics of the process are *hopelessly* obscured by the district-distinguishing roles
of the lagged variables.

vant to policy decisions in the way indicated at the outset. The size-of-discovery variable used in the present analysis is one whose value is not known with certainty for at least several years after the discovery is made. This reflects the fact that only a very small part of the oil in newly discovered fields is available for production in the year of discovery. The process which transforms estimates of discoveries into proved reserves is a long one, entailing a far greater capital investment than does the drilling of a wildcat well, although a somewhat less risky venture is involved. This analysis has applied strictly to wildcat drilling and to total amounts of oil and gas in known fields. This is not the same as performing the more directly relevant analysis of the responses to economic incentives of other exploratory and development drilling and of known and readily available oil and gas reserves.[51]

As is clearly the case with wildcat drilling, development drilling is doubtless more sensitive to economic incentives in small fields than in big ones. It is the development of *marginal* fields that is curtailed when prices fall. It follows as in the case of the discovery of new fields, that the price elasticity of available oil reserves is probably substantially less than the price elasticity of the drilling and other expenditures which produce those reserves. It seems reasonable to suppose that the effect is roughly of the same order of magnitude as that found in our results, but this is only a highly plausible conjecture and must remain so in the present state of industry statistics.[52]

It follows, then, that the best available estimate of the elasticities of exploration and development drilling as a whole and of available oil with respect to crude price—and hence with respect to any other factor that affects profits in the same way, such as the depletion allowance—places the drilling elasticity in the neighborhood of $+2.8$ and the reserves elasticity at considerably less, say, $+0.3$ (subject to the reservations already discussed). What is to be concluded from this as to U.S. oil policies and as to the worth of a major argument which has been offered in support of those policies that there is a high degree of sensitivity in this area?

[51] Of course, such considerations also open the question of whether readily available oil reserves are themselves the really relevant variable for national defense. On this, see S. L. McDonald, *Federal Tax Treatment of Income from Oil and Gas* (Washington: The Brookings Institution, 1963), pp. 85–86.

[52] There is simply no way of performing an analysis similar to the present one on development drilling and its results. There are no statistics on such drilling or on developed reserves organized by year of field discovery. A highly useful start in this direction would be the continual updating of the National Petroleum Council study used in the present work, as this would provide a yearly record of additions to known fields. Similarly, it would be desirable if the American Petroleum Institute and the American Gas Association were to report their extensions separately, and were to give the reasons for revisions. None of these suggestions could be implemented without a fair amount of work, and no such work would be free of arbitrary elements. Nevertheless, the work involved would be far less than that involved in separating past data, as was done in the National Petroleum Council study, and the arbitrary decisions could be made taking that study as a base for some of them. In this way, a useful record of the industry's performance and of the nation's acquisition of available oil and gas could be built up over the next several years.

These recommendations were also considered at a seminar of which the author was a member. See Lovejoy and Homan with Galvin, *op. cit.*, pp. 105–16.

Before answering this question, let us look critically at the evidence marshalled here. The numbers serve as indications of magnitude only. Our results—as is invariably and inevitably the case in an analysis of this nature—refer to a particular period and to a particular set of values of the variables involved. It has been argued in several places that the relationships may well have changed in various ways since 1955; they may also have changed in others. Moreover, the sensitivities of the dependent variables to economic incentives may be quite different from those here found when those incentives are drastically cut or drastically increased. It also has been emphasized, in discussing the effect of shutdown days, that it is important to consider the means as well as the ends of an incentive policy, as the means may themselves be incentives.

Furthermore, the definition of what is a "high" degree of sensitivity depends upon several factors that are outside the context of this analysis: the reason for the question being asked, for example; the policy objectives in view; and the larger framework involved. While elasticities greater or less than unity provide a convenient terminology for "high" or "low" sensitivities, that is only a terminology. An elasticity of $+2.8$ may be "low" for some purposes and an elasticity of $+0.3$ intolerably "high" for others. The present study is but an aid in this regard.

Nevertheless, when all is said and done, our findings do cast considerable doubt on policy arguments resting on the assertion that oil discoveries are highly sensitive to economic factors. To the extent that this argument depends on evidence as to the high sensitivity of wildcat drilling—or other exploratory and development drilling—it substantially overstates the case. The sensitivity to economic forces of oil (and gas) discoveries is far less than the apparent sensitivity of industry expenditures. If an acceptably large stock of domestic oil in place is being maintained because discoveries are acceptably high, then our results indicate that this would not change much if special incentives were reduced. If, on the other hand, a satisfactory stock is not being maintained, it may require a great increase in such incentives to secure the desired discovery increase. Whether or not such an increase in incentives would have to be so large as not to be worth the result is clearly outside the province of this study. In any case, it would seem that some or all of the special policies adopted by the United States government are open to question if their sole justification lies in the premise that discoveries react strongly to economic incentives. Whether these policies are justified on other grounds requires more detailed evidence and is a question that this analysis does not attempt to answer.

II

Measuring the Effects of Depth and Technological Change on Drilling Costs

1. Introduction and Summary

That the extractive industries incur inevitably increasing costs through time as the richer deposits become exhausted is an assumption that is being challenged by many who view the continual advances of technology as a counterbalance to such a tendency.[1] If this view is correct, it has particular implications for the petroleum industry, where costs and cost trends not only act as determinants of current investment but also are significant factors in the policy decisions made by regulatory bodies, especially those pertaining to rate setting.[2] This study tests out the behavior of one link in the chain of economic costs of petroleum production.

The uncertainties inherent in the petroleum industry and the long and variable time span for production of crude petroleum from a given field render somewhat hazardous any evaluation of economic costs across the entire process of discovery, development, and production. In addition, the available data covering all segments of the process are quite insufficient for rigorous analysis.[3] Therefore, to test the application to petroleum of the

---

[1] See, for example, O. C. Herfindahl, *Copper Costs and Prices, 1870–1957* (Baltimore: Johns Hopkins Press, 1959); and B. C. Netschert, *The Future Supply of Oil and Gas* (Baltimore: Johns Hopkins Press, 1958).

[2] In the petroleum industry, these are principally natural gas rates set by the Federal Power Commission.

[3] See J. E. Hodges and H. B. Steele, *An Investigation of the Problems of Cost Determination for the Discovery, Development, and Production of Liquid Hydrocarbon and Natural Gas Resources,* The Rice Institute Pamphlet (Houston), XLVI, No. 3 (October, 1959); and W. F. Lovejoy and P. T. Homan with C. O. Galvin, "A Study of the Problems of Cost Analysis in the Petroleum Industry," *Journal of the Graduate Research Center* (Southern Methodist University), XXXI, Nos. 1 and 2 (February, 1963).

theory that advancing technology tends to offset advancing cost, it seems practical to limit analysis to one activity—the drilling of petroleum wells—on the costs of which a great deal of information is available.

Some of even these data are open to question (see Sections 4 and 6). One must also bear in mind that drilling cost is only one large item in the total costs of producing a barrel of oil or a million cubic feet of natural gas. Nevertheless, it is believed that an analysis of drilling costs can provide some indication of the effect of well depth and technological change on U.S. petroleum costs in general.[4]

The analysis set out in the following pages produces considerable evidence of reductions in drilling costs during 1955–56–59, the period studied, despite the trend toward drilling deeper and thus more costly holes. The tendency toward cost decreases is more apparent in the case of dry wells than of productive wells and is stronger for shallower wells than for those of great depth. It seems highly probable, therefore, that in most areas the real costs of drilling per well (as opposed to the effects of price level changes) fell during the period because costs of increases in depth per well were more than compensated for by cost reductions brought about by increased efficiency.

Among the major technological advances of around this time were the chert bit, air and gas drilling, and slim-hole drilling. The chert bit caused a dramatic drop in the cost of boring very hard and abrasive strata such as those found in parts of Texas, Oklahoma, New Mexico, and in the Rocky Mountains. Gas and air drilling likewise brought significant cost reductions in favorable conditions, although in certain soft formations of the Gulf Coast area these processes could not be used. Slim-hole drilling is relatively unattractive for wells which are confidently expected to be producers, but the relative cheapness and portability of the process make it very suitable for wildcatting.

Such innovations as these have undoubtedly helped to decrease over-all drilling costs; but they also point up the need for caution in analyzing data drawn from areas of widely differing geologic formation. To the extent possible, allowance for such differences has been made in this study.

A basic finding of the study is that, letting Y be per-well costs in dollars (excluding set-up and other overhead costs), X be depth in feet, α and K constant parameters, and e the base of the natural system of logarithms, the relation

$$Y = K(e^{\alpha X} - 1)$$

gives an excellent fit to the available data for practically any state or subdivision in terms of measuring the marginal and total costs of drilling. ("Marginal cost" in this context may be defined as the difference between

[4] Drilling costs have figured importantly in hearings before the FPC. See Federal Power Commission, *In the Matter of: AREA RATE PROCEEDING, et al. (Permian Basin Area)*, Docket No. AR 61–1, *et al.* (hereafter referred to as *Permian Basin Proceeding*).

the cost of drilling a well of X feet and a well of $X + 1$ feet.) This function agrees with the data in showing that costs per well rise rapidly as depth increases and that the rate of such rise also increases with increased depth. It is also supported by some theoretical considerations.

The data on which the study draws are basically the statistics for 1955–56 and 1959 published in the *Joint Association Survey of Industry Drilling Costs (JAS)* and the sample from which the 1959 published figures were constructed. The reservations concerning these, particularly the published figures for 1959, are fully discussed in later pages, as are the steps taken to make the sets of figures comparable over time. When estimates for 1959 are made on a sample basis and when these are substituted for the estimates from the 1959 published figures, the conclusions of this study concerning changes over time remain essentially unchanged. Here in this summary, it is sufficient to point out that the published figures for 1959 should be approached with caution. They tend to overstate both marginal and total drilling costs for shallow-depth wells—the great majority—and to understate them for deep wells. The figures become approximately correct at a depth of 10,000 feet, whereas the average well is less than 5,000 feet.

Apart from inconsistencies in the derivation of the statistics from one period of time to another, there are other ways in which changes in the cost-depth relationship can occur. For example, a given technological change might be of great advantage in shallow well drilling but of no benefit in deep well work. Another technological change might work in exactly the opposite manner and, in fact, it is seldom that significant technical advances benefit all types of wells equally. Or the exploitation of a previously untouched area may disturb pre-existing relationships.

Even allowing for the limitations of the information available, it seems inescapable that if domestic petroleum industry costs increased in the period 1955–56—1959 (and this is by no means certain), the causes are to be found in price level changes, or discovery, development, and production costs, not in drilling costs. For the industry as a whole, real drilling costs declined during the period, and while it is impossible to be sure that future technological advances will maintain this trend, there is no evidence that drilling costs must inevitably rise as wells get deeper, even though the statistics for any given year may show a sharp increase of costs with depth.

2. The Model

The focus of this study is on the relation between drilling cost and depth —controlling insofar as possible for differences in geology—and on the way that technological change has acted to alter that relation. Let us then consider what form that relation may be expected to have.

A glance at the data for any state or state subdivision shows immediately that costs per well are a rapidly rising function of depth and that the rate of rise is also a sharply increasing function. How, then, might the marginal costs vary with depth? What we are measuring here, it should be pointed out, is not the cost of drilling an extra foot in a given well, but the difference, with given technology, in the cost of two wells, one of which is designed to be a foot deeper than the other—a long-run marginal cost.

The discussion which follows was suggested by the data for a few principal states. It is intended only to be suggestive and to lend some theoretical plausibility to the functional form adopted. The true test of that form is its application in the nearly four hundred regressions performed and reported later in this study. It will be seen that the test is more than passed on whatever data it is performed.

Two families of hypotheses immediately come to mind. The first of these is that marginal costs are some direct function of depth. Thus, let Y be those costs which vary with depth (no overhead costs), and X be depth itself, we might write:

$$(1) \qquad\qquad dY/dX = F(X).$$

While this is, of course, formally correct, it is not very easy directly to specify the form of the function $F(X)$. Moreover, there is a sense in which depth in feet is not a proper measure of the variables which affect marginal costs.

Depth affects marginal costs in several ways through a variety of different factors. Thus, for example, the rise of temperature with depth, among other factors, increases the probability that a bit will have to be replaced an additional time in a well drilled an additional hundred feet; the amount of time that such replacement takes goes up with depth; the mechanical energy lost in the drilling process goes up, and so forth. In addition, there are items in marginal costs that increase rather slowly, if at all, with depth: principally, the additional costs of certain materials required to drill a deeper well. Depth affects each of these factors rather differently and to specify such effects separately and precisely would require a very detailed and complicated function.

This suggests the following. What is required is a measure of the increased difficulty of drilling a given amount of additional footage at a given initial depth; an obvious measure of this is the cost of drilling to the given initial depth itself. All the factors which increase marginal cost as depth goes up are operative throughout the drilling of a well and the sum of their effects is the variable cost of drilling itself. In this sense, what is important in determining the cost of the next hundred feet is not how far down a well has been drilled, but how much it has cost to get that far, how difficult the going has been. Physical depth is not a particularly good measure of "economic depth," but cost itself is.[5] This suggests that we ought to write instead of equation (1):

(2) $$dY/dX = G(Y),$$

and our problem now becomes the specification of the function $G(Y)$.

Fortunately, just about the simplest choice of $G(Y)$ appears to work extremely well in the results. Clearly, since Y must be zero when X is zero (Y includes only costs which vary with depth), dY/dX cannot be simply proportional to Y, as marginal cost does not approach zero as we consider ever more shallow wells. On the other hand, there is no reason why such proportionality cannot represent the relation of marginal to total variable cost as depth becomes very great. (This would be a plausible—but by no means a necessary—consequence of the argument as to costs themselves as the best measure of the effect of depth on marginal costs.) We thus write:

(3) $$dY/dX = H + \alpha Y$$

where $H > 0$ is the limit of marginal cost as depth goes to zero and $\alpha > 0$ is a constant parameter whose interpretation will be discussed below.

Rewriting:

(4) $$\frac{d(Y + K)}{dX} = \alpha(Y + K)$$

where $K = H/\alpha$. This yields:[6]

(5) $$d \log (Y + K) = \alpha dX$$

which, on integration becomes:

(6) $$\log (Y + X) = \alpha X + \log C$$

where C is an arbitrary constant. Taking anti-logarithms and rearranging:

(7) $$Y = Ce^{\alpha X} - K.$$

By definition, however, Y is zero when X is, so that:

(8) $$C = K,$$

and the final form of the function is:

(9) $$Y = K(e^{\alpha X} - 1).$$

This is the function to be fitted to the data. It clearly makes no allow-

---

[5] Since cost is itself a function of depth, depth will in fact enter into the determination of marginal cost, so that equation (1) is formally correct as stated.

[6] All logarithms are natural.

ance for overhead costs; however, when it is fitted with an additional constant term for such costs, that constant term turns out to be zero to all significant digits in every one of the many cases tried, showing that overhead costs are not included in the data.

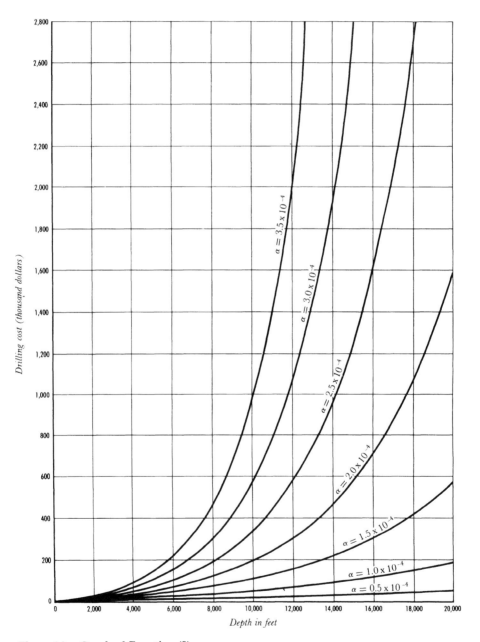

Figure 1A.   Graph of Equation (9):
    $K = 30,000$, $\alpha = 0.5 \times 10^{-4}$ to 3.5×10^{-4}.

What are the properties of the function (9), and how are we to interpret its parameters? Study of the function as graphed on arithmetic and semi-logarithmic paper for selected values of the parameters (Figures 1A–1D) may aid in following the discussion. By construction, the function goes

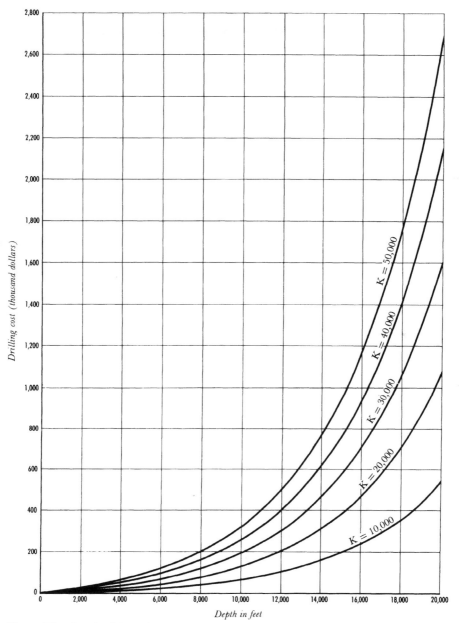

Figure 1B.   Graph of Equation (9):
    $\alpha = 2.0 \times 10^{-4}$, $K = 10,000$ to $50,000$.

through the origin and satisfies equation (3). It follows that one way of interpreting α (which has the dimension of one over depth) is as the asymptotic ratio of marginal to total cost[7] as depth becomes very large. There are other interpretations, however.

From equation (9) we have:

(10) $$dY/dX = \alpha K e^{\alpha X} = H e^{\alpha X} .$$

Recall that H is the limit of marginal cost as depth approaches zero, then:

(11) $$\frac{(dY/dX)}{H} = e^{\alpha X}$$

or

(12) $$\frac{d \log (dY/dX)}{dX} = \alpha .$$

α may thus be thought of as the percentage change in marginal cost induced by a unit increase in depth;[8] it is a measure of the curvature of the depth-cost relationship—the greater is α, the greater the curvature.[9] It is important to note that treating this as a constant parameter gives an exceptionally good fit to the data in essentially every case; this implies that at all depths the same absolute depth increase has the same multiplicative effect on marginal cost.[10] As will be evident in the results, that multiplicative effect is sizable in many areas.

It is thus possible to use α to determine the absolute depth increase which would increase marginal cost by a given per cent. This can also be done (although not as neatly) for total cost. Thus, let X^1 be any non-zero initial depth; let X^2 be that depth at which cost is twice as great as cost at X^1. Then, from equation (9):

(13) $$K(e^{\alpha X^2} - 1) = 2K(e^{\alpha X^1} - 1)$$

[7] The term "variable" is dropped in referring to Y, it being understood that fixed costs are not included in this analysis.

[8] The author is indebted to O. C. Herfindahl for pointing this out.

[9] When equation (9) is estimated by least squares, as described in Appendix A, and K is therefore chosen to satisfy the appropriate minimizing equations, it is easy to see that: $K = \Sigma Y(e^{\alpha X} - 1)/\Sigma(e^{\alpha X} - 1)^2$. It follows that K approaches infinity as α approaches zero. It can then be shown by L'Hôpital's Rule that as α approaches zero, the function (9) approaches a straight line through the origin, so that α is indeed a measure of curvature even in the limit.

[10] The *Joint Association Survey of Industry Drilling Costs, 1959* (New York: American Petroleum Institute, Independent Petroleum Association of America, and Mid-Continent Oil & Gas Association), remarks that "as wells go deeper into the earth, drilling costs increase approximately geometrically" (p. 5). This is nearly correct for total drilling costs—as shown from the good fit obtained from equation (9)—but is much more accurate when applied to marginal cost, as shown in equation (11).

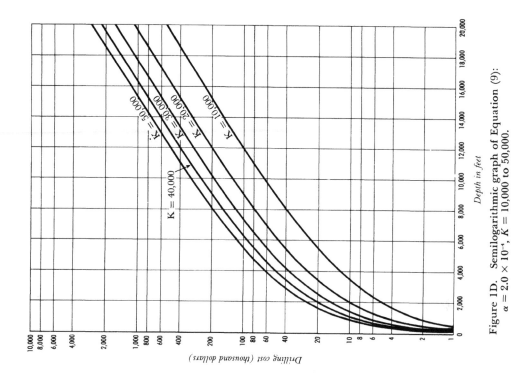

Figure 1C. Semilogarithmic graph of Equation (9):
$K = 30,000$, $\alpha = 0.5 \times 10^{-4}$ to 3.5×10^{-4}.

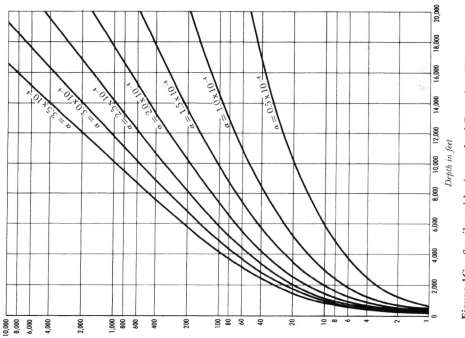

Figure 1D. Semilogarithmic graph of Equation (9):
$\alpha = 2.0 \times 10^{-4}$, $K = 10,000$ to $50,000$.

49

whence:

(14) $$(X^2 - X^1) = (1/\alpha) \log (2 - e^{-\alpha X^1}).$$

As the initial depth increases, this approaches:

(15) $$\text{Lim } (X^2 - X^1) = (1/\alpha) \log 2$$
$$X^1 \rightarrow \infty$$

which can also be shown to be the expression at all initial depths for the absolute depth increase in which marginal cost is doubled. Since $e^{-\alpha X^1} > 0$, the exact expression in equation (14) is always less than its limit in (15). This is quite important, since the results (see Figure 2 and Appendix B, Table B2) show that cost doubles asymptotically (and marginal cost doubles

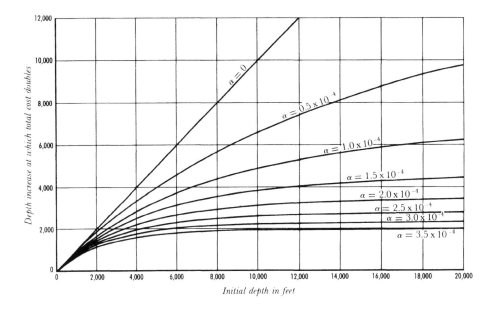

Figure 2. Depth increase at which total cost doubles as a function of initial depth: $\alpha = 0.5 \times 10^{-4}$ to 3.5×10^{-4}.

at all initial depths) in under 3,500 feet for a good many areas, the figure being substantially less for some (and more for others). As stated, the approach to this limit is from below, so that costs at actual initial depths double even faster than this would indicate.

Turning now to the interpretation of K: this coefficient, which has the dimension of dollars, is not a fundamental parameter of our model; rather it is derived as the ratio of H to α. It is not surprising, therefore, to find that

K appears to have no very simple interpretation in its own right, so long as discussion is restricted to the function (9) as it applies to a single area, type of well, and point of time. It is easy to interpret *changes* in K over time or differences in K over areas, however.

To begin with changes over time: Suppose that the prices paid by operators and the contractor rates, or both, change in such a way that the costs of drilling a well to any depth are changed in proportion to their original values. This will show up in our analysis as a change in K with α held constant. Similarly, suppose a change in drilling technology is introduced which has the effect of reducing costs at all depths by the same proportion— a change which will be termed "depth-neutral." The result will be to reduce K in the same proportion but again to leave α unaltered. Changes in K over time with a constant α, then, represent price and technology changes which are neutral with respect to depth in that they affect all wells in proportion to their costs. (Such changes shift the function upward by a constant amount when graphed on semilogarithmic paper—see Figure 1D.)

Similarly, differences in K over regions with the same α stand for differences in drilling difficulty which are depth-neutral in the same sense. An analogous statement holds for differences in type of well.

Of course, changes over time (or differences over regions) are not likely to be depth-neutral in the above sense. Technological change in particular is likely to reduce costs proportionately more in deep wells than in relatively shallow ones or vice versa, rather than to affect both types of wells equally. (In the former case, the changes will be referred to as "depth-favoring," and in the latter as "shallowness-favoring.") This suggests that it is important to discuss changes in α.

Suppose, then, that K and α both change. Let α_0, K_0, and H_0 be the parameters before the change and α_1, K_1, and H_1 be the parameters after it. The resulting ratio of marginal cost after the change (at any depth, X) to the corresponding value before the change is (where $MC_1(X)$ and $MC_2(X)$ are the two marginal cost functions):

$$(16) \qquad \frac{MC_1(X)}{MC_0(X)} = \frac{H_1 e^{\alpha_1 X}}{H_0 e^{\alpha_0 X}} = (H_1/H_0)\, e^{(\alpha_1 - \alpha_0)X}.$$

It follows that:

$$(17) \qquad \frac{d\log(MC_1(X)/MC_0(X))}{dX} = \alpha_1 - \alpha_0,$$

so that the change in α measures the effect of depth on the effect of the change on marginal cost. Thus, an increase in α indicates that costs were reduced less (or increased more) by the change in deep wells than in shallow ones, while a decrease indicates the opposite effect. Precisely, the change in

α measures the percentage increase per unit depth in the effects of the change on marginal cost.

To a first approximation, this is also true of the effects on total cost itself. Let $Y_1(X)$ be total cost after, and $Y_0(X)$ be total cost before, the change:

$$(18) \qquad \frac{d \log (Y_1(X)/Y_0(X))}{dX} = (\alpha_1 - \alpha_0) + \frac{\alpha_1}{e^{\alpha_1 X} - 1} - \frac{\alpha_0}{e^{\alpha_0 X} - 1} .$$

Denote the expression on the right by Z and observe that:

$$(19) \qquad \operatorname*{Lim}_{X \to \infty} Z = \alpha_1 - \alpha_0 ,$$

so that for very deep wells, $\alpha_1 - \alpha_0$ is a good approximation to the effects of depth on the effects of the change in question.[11] It can be shown in the following way that the precise value of Z always lies between zero and $(\alpha_1 - \alpha_0)$:

To see that the sign of Z is the same as that of $(\alpha_1 - \alpha_0)$, note that if $\alpha_1 = \alpha_0$, $Z = 0$. It therefore suffices to show that Z is monotonically increasing in α_1. Differentiating, we have:

$$(20) \qquad \partial Z / \partial \alpha_1 = \frac{(e^{\alpha_1 X}) (e^{\alpha_1 X} - 1 - \alpha_1 X)}{(e^{\alpha_1 X} - 1)^2} .$$

The denominator is obviously positive, as is the first factor in the numerator. The second factor in the numerator is also positive for any positive X as can be seen by expanding $e^{\alpha_1 X}$ in Maclaurin Series. Hence $(\partial Z / \partial \alpha_1) > 0$ and the sign of Z must be the same as that of $(\alpha_1 - \alpha_0)$.

To see that Z is always less in absolute value than $(\alpha_1 - \alpha_0)$, it suffices to show that the sign of the sum of the last two terms in Z is opposite to that of $(\alpha_1 - \alpha_0)$. First, suppose that $(\alpha_1 - \alpha_0) > 0$, then for any $X > 0$:

$$(21) \qquad \frac{\alpha_1}{e^{\alpha_1 X} - 1} = \frac{\alpha_1}{\alpha_1 X + (\tfrac{1}{2})(\alpha_1 X)^2 + \cdots}$$

$$= \frac{1}{X + (\tfrac{1}{2})\alpha_1 X^2 + \cdots} < \frac{1}{X + (\tfrac{1}{2})\alpha_0 X^2 + \cdots}$$

$$= \frac{\alpha_0}{\alpha_0 X + (\tfrac{1}{2})(\alpha_0 X)^2 + \cdots} = \frac{\alpha_0}{e^{\alpha_0 X} - 1} ,$$

while if $(\alpha_1 - \alpha_0) < 0$, the inequality sign is reversed. This demonstrates the desired result.

It follows that the sign of $(\alpha_1 - \alpha_0)$ indicates the depth-affecting nature of the change, a depth-favoring change having a negative sign and a shallowness-favoring change a positive sign. The magnitude of $(\alpha_1 - \alpha_0)$ itself gives an exact measure of the percentage effect of a unit depth increase

---

[11] Note the parallel to the earlier discussion of equations (13) to (15).

on the effects of the change on marginal cost, and an upper limit (which is approached asymptotically) for the percentage effect of a unit depth increase on the effects of the change on total cost.

To sum up, to the extent that a change affects the costs of wells at all depths proportionately, it appears as a change in K; systematic deviations from proportionality with depth appear as a change in α. Analogous statements hold when comparing cost functions for different regions or for different types of wells.[12]

3. Forces Altering the Cost Function over Time

We must now ask what we should expect to see when the function (9) is fitted separately to data for the same region and type of well for different years. There are several forces which might act to change the parameters over time.

The simplest of these forces is a change in the prices of goods and services purchased by operators in order to drill a well. To the extent that such a change affects all wells in proportion to their cost, this will result in a change in K, as has just been shown, and not in a change in α (the rate at which marginal cost increases with depth). To the same extent, therefore, it is possible to remove such a purely monetary effect through adjustment by a price index, and this we shall attempt to do.

On the other hand, there is no reason to expect the effect of such price changes to affect wells at different depths proportionately. The items which make up any such price index do not all change by the same proportion, and it is reasonable to suppose that wells drilled to different depths use the various goods and services involved in differing proportions. Therefore α may well be changed by such price changes, as their effect on costs may not be depth-neutral.

Second, in the absence of technological change, there is probably a natural tendency for H (the limit of marginal cost as depth goes to zero) and α to increase. This is so because, other things being equal, wells will be drilled in low-cost areas earlier than in high-cost ones. Of course, other things may not be equal—the amount of oil to be found in low- versus high-cost areas, for example; nevertheless, that such a tendency clearly exists for the United States as a whole is shown by the fact that our results (see Tables 1A–1D) reveal a strong negative correlation over different areas between H and α. This means that areas with relatively high costs for shallow wells tend to have relatively low rates of increase of cost with depth and vice versa, while

---

[12] A shallowness-favoring change which reduces K and raises α may well *raise* marginal cost above some depth while lowering total cost for all wells below a much greater depth which is beyond drilling range.

areas with both relatively high α and high H tend not to be drilled.[13] This indicates that there is a tendency, in the United States as a whole, to drill low-cost before high-cost areas and hence for α and H to increase over time in the absence of technological change. It seems unlikely that the available segregation of the data into states and subdivisions thereof is fine enough to eliminate this effect within such data areas.

This argument has a further consequence. Clearly, the tendency just discussed must be somewhat stronger in the drilling of exploratory wells than in the drilling of more or less proved structures, since the shift toward high H and α areas clearly begins with the drilling of new fields while the drilling of old ones is the drilling of an accumulation of old areas. Since exploratory wells have a much higher proportion of dry holes than do other wells, the effect in question should certainly be more pronounced in the drilling of dry holes than in the drilling of productive wells, though present in both. Technological change may partly or wholly offset this effect in both types of wells, but we should still expect to see a greater increase (lesser decrease) over time in H and α in the fitted curves than would be the case if geology were fully controlled in the data, unless the new techniques are first used in exploratory drilling.

Technological change is clearly a factor of considerable importance, for a number of changes in drilling technology occurred in the late 1950's. It is the only factor, moreover, which obviously directly acts to reduce costs. But the effects of technological change on our results must be interpreted with some care, for it is a fact that some changes in technology, for example those that on their face appear to be depth-neutral or even depth-favoring, do not show up as such in the data available.

Consider, as an illustration, the introduction of the chert bit.[14] This was an innovation which made economically feasible the drilling of extremely hard and abrasive formations, such as occur in West Texas and New Mexico or Oklahoma and the Rocky Mountains. Since the cost of drilling such formations dropped dramatically, the chert bit would appear to be a depth-neutral innovation which acted to reduce K and leave α unaffected. This was probably the case for the drilling-cost function applying to wells drilled in chert or similar formations. However, the available data for the regions studied are not broken down by type of formation. Further, the fact that the chert bit made hard and abrasive formations economically feasible for drilling encouraged a relative shift toward drilling such formations in the areas where they occur. Since it is entirely possible that drilling hard formations with the chert bit costs more than drilling other formations without it and that the rate of cost increase with depth is also greater, the net effect on drilling costs in the data area as a whole may well have been

---

[13] Areas with relatively low α and low H are observed, of course; however, not very many show up, presumably because such areas have been drilled extensively in the past.

[14] "Drilling Practices," *World Oil*, Vol. 148, No. 1, January, 1959, p. 164.

to *increase* K or α or both, even though the effect on the costs of drilling the hard formations themselves was doubtless the reverse.

To put it another way, the introduction of the chert bit is an example of a change in drilling technology which accelerated the movement to more costly areas. Despite the fact that the technological change itself was cost-reducing, the net effect on average drilling costs in the area for which data are separately available may well have been cost-increasing.

Further, if wells drilled on chert formations are not distributed over different depths in the same proportions as are wells drilled on the other formations in the same data areas—because of differences in the depths at which oil or gas occurs, for example—then our estimate of α may shift as a result of the introduction of the chert bit, even if that introduction was a depth-neutral technological change as regards the drilling of chert formations themselves.[15]

The above example works as it does because the technological change involved, the chert bit, had effects which were different in different types of formations. Another innovation introduced at about the same time was gas or air drilling.[16] This type of drilling reduces costs and increases penetration rates, but cannot be employed in certain soft formations such as occur principally along the Gulf Coast. It follows that its introduction doubtless prompted a *relative* shift in the Gulf Coast areas away from such formations, with results for our cost functions as estimated that might conceivably be in any direction. However, to the extent that air and gas drilling was introduced (at least at first) principally in relatively shallow wells because of the increased danger of blow-outs at depth, the result will clearly be an increase in α in a function fitted to a cross-section of wells for any area in which this occurred.

While the above examples both turn on differential effects of technological change in different types of formations, it is important to realize that only the sharpness of the effects noted is due to this property. Even if a technological change reduced costs equally in all formations, it would prompt, other things being equal, an acceleration of the relative shift to more costly formations. Its effects would therefore tend to show up in the reported data as, at least, less than its true cost-reducing effects.

One other technological change has probably affected the costs of dry holes more than those of productive wells. This is the introduction of slim-hole drilling.[17] Aside from any possible difference in completion costs as

---

[15] There is some evidence in the results for the affected areas that the introduction of the chert bit did indeed have the net effect of increasing α and increasing total costs somewhat in some depth ranges; however, the example is meant as a plausible illustration of a possible effect rather than as a statement of historical fact.

[16] See, for example, James T. Morris and Robert P. Ramsey, "When Does It Pay to Drill with Air?" *World Oil*, Vol. 142, No. 1, January, 1956, pp. 112–22.

[17] See R. W. Scott and Jack F. Earl, "Small Diameter Well Completions," *World Oil*, Vol. 153, No. 2, August, 1961, pp. 57–66.

such in dry and productive wells, the portability of the equipment and the other properties of slim-hole drilling make it especially attractive for wild-cat drilling. Furthermore, the disadvantages of a slim-hole completion as regards lifting costs make it relatively unattractive for wells that are con-fidently expected to be productive. Since wildcats account for a much greater share of dry holes than of productive wells, the effect is thus a greater use of slim-hole drilling in wells which turn out to be dry holes than in oil or gas wells. Up to a point, this may act to offset the greater rate of movement of dry holes to high-cost areas noted above.

This by no means exhausts the list of changes in drilling technology which have been introduced in relatively recent years and, in balance, it is hard to say what the effects on our function as fitted should be. Aside from the effects already discussed, however, one thing is clear: Since costs rise so very sharply with depth, there is a natural tendency to look for innovations which will be especially effective at depth and thus reduce α. Whether the result of the technological changes that actually took place were in fact the reduction of costs at depth *relative to the costs of shallower wells* in the reported data when all the effects we have discussed are taken into account, is another matter.

We turn, then, to the last principal factor which might change our parameters as estimated, the nature of the data themselves. The data to which our function will be fitted are a cross-section of wells drilled in a given area.[18] Aside from the problems already raised, this brings up the problem of the homogeneity of the data for a given year. If formations within a given data area vary in their costliness, and if the sample (or the population) of wells from that area contains relatively more deep wells from the low-cost formations, for example, and relatively more shallow wells from the high-cost ones, the result will be a relatively low value of α as estimated for that data area. If, on the other hand, there are relatively more deep wells from the high-cost formations and relatively more shallow wells from the low-cost ones,[19] then α as estimated will be high. The estimated value of α will be different, then, for different mixes of wells from different formations; K will also be affected. Among other things, it follows that a change over time in the depth-formation relation, due to technological change as suggested above or to any other factor, can change our estimates even though the functions within each homogeneous geological area remain the same.

This sort of problem is, of course, the basic reason for collecting data by relatively small areas. It is even more serious as regards estimates for the

[18] More precisely, the observations are on depth-class means rather than on individual wells and a weighted regression procedure is used. (See the discussion in Appendix A.) This does not affect the present discussion.

[19] "High" and "low" cost formations are referred to in the sense of comparing drilling costs at a given depth. Of course, a formation which is "high-cost" to one depth may be "low-cost" to another; we are simplifying for the sake of the exposition. The general point being made does not depend on the particular features of the example.

United States as a whole. The problem is not restricted to procedure but is endemic in any analysis of the data themselves.

There is one sense, however, in which this problem is less serious than it may appear. Since formations in a given area differ not only in their costliness but also in the depths at which oil is to be found, then, in a meaningful sense, the cost of drilling for oil to a given depth in that area *is* measured by the cost of wells actually drilled to that depth. If an operator decides to sink a deeper well, that decision will be accompanied by a shift in his operations to another formation. The relevant function for decision is thus not the cost function for wells in a given formation, but the cost function across formations, which is what we estimate. A shift in the parameters of that function over time due to the effect under discussion is thus not necessarily irrelevant, as it may reflect a real shift in drilling costs as seen by operators. The problem raised by this effect is therefore partially alleviated.

Unfortunately, such alleviation is only partial. Aside from the fact that it would be desirable to isolate the effect just noted from that of a change in the drilling cost function on given formations, the argument just given is valid only when the depth-formation relationship in the sample of wells from which the available data are constructed is the same as that for all wells in the area studied. If the sample is not representative in this respect, but is biased toward one or another formation at different depths, the across-formation relationship estimated will not be the appropriate one. It follows that changes in the representativeness of the *sample* alone can change our parameters, and this is clearly undesirable.

It is necessary, therefore, to examine the nature of the data and also the way in which these data are constructed, for it is evident that the construction procedure as well, given the sample on which the published data are based, can affect our estimated parameters and their behavior over time.

4. The Joint Association Survey of Industry Drilling Costs

In 1953, 1955 and 1956, and 1959, the American Petroleum Institute, the Mid-Continent Oil & Gas Association, and the Independent Petroleum Association of America conducted surveys of drilling costs. These surveys, known as the *Joint Association Survey of Industry Drilling Costs (JAS)*[20] present detailed data on drilling costs and depth by depth-class averages for wells drilled by states and by subdivisions of some states.[21] The data are broken down in the 1955–56 survey by dry holes and productive wells, and

---

[20] Available from the organizations mentioned. The author, however, was unable to persuade the *JAS* to provide him with the original sample data.

[21] The Appalachian area includes New York, Ohio, Pennsylvania, and West Virginia and is presented as a unit.

in the 1959 survey by dry holes, oil wells, and gas wells,[22] but are not broken down by well type in the 1953 survey. Only the later two surveys, therefore, are really comparable.

The *JAS* procedure for the 1959 survey was essentially as follows:[23] A questionnaire was distributed to as many companies as possible and the replies received (after minor adjustments for obvious inconsistencies of reporting) constituted the sample. The sample data for a given data area were then blown up to represent the entire population of wells drilled in that area (as reported in the *Oil and Gas Journal*) in a series of steps. The procedure used need not concern us as regards the estimates of the number of wells and the footage drilled.

The technique used for the cost data themselves does concern us, however. The sample data were organized into total footage and total cost for all wells in each of nine depth ranges, and cost per foot was plotted for the resulting points against depth. A freehand curve was then drawn through the results and the estimated costs per foot of the unreported wells were read off using depth figures previously constructed.[24] The costs for such wells were then added to reported costs in the depth class and only the totals as so constructed for each depth class were published.

It is obvious that there are three important questions which must be considered in evaluating the *JAS* for our, or any, purposes:

A) In what sense, if any, is the *JAS* sample representative of the population of wells drilled?

B) How well does the organization of the sample data by depth class totals or averages represent the sample data themselves?

C) How valid and reliable are the procedures used in blowing up the sample to secure estimates of population figures?

In discussing these questions, the primary interest will be in their implications for the use of the *JAS* in a study such as this. Some comment on wider issues will also be made, however, and this leads to our final question:

D) How can the *JAS* be improved?

[22] For productive wells the costs are further broken down into tangible and intangible costs. (The costs include those of equipping the well through the control manifold known as the Christmas tree.) The author has made no use of the latter breakdown but has dealt exclusively with total cost since extremely good fits were obtained thereto. The 1959 survey also gives a somewhat finer breakdown by area than do the earlier surveys.

[23] For a more detailed account, see the *JAS*, 1959, pp. 10–14. See also *Permian Basin Proceeding*, pp. 3985 ff. (cross-examination of R. M. Brackbill).

[24] This was not done with the 1955–56 survey, the sample figures for each depth class being used whenever the sample was deemed adequate in size and the cost figures not clearly out of line (letter of David V. Hudson, Jr., to the author). How often such exceptions occurred or what was done when they did is not known to the author, but it is fair to say that the strictures on freehand curve drawing given below apply to the 1959 survey rather than to that of 1955–56. It follows that the 1955–56 survey probably suffers from considerably less bias than our results show to be present in the published 1959 figures. The data are not available with which to check this, however.

The practice, evidently followed in all the surveys, of deleting outlying observations is not without hazard. It may result in hiding just those irregularities in nature which a survey ought to discover. (Cf. M. Nerlove, *The Dynamics of Supply: Estimation of Farmers' Response to Price* [Baltimore: Johns Hopkins Press, 1958], pp. 98–99.)

A) In what sense, if any, is the JAS *sample representative of the population of wells drilled?*

The *JAS* sample is not a random sample in any statistical sense.[25] Moreover, it is not a representative one, since it tends to overrepresent deep wells.[26] For analytical purposes this is unfortunate, since it detracts from the argument given above as to the appropriateness of data from different formations in the same area. In addition, as the relative weights given to wells drilled on different formations doubtless change from survey to survey, any comparison of cost functions over time or over different states is rendered rather dubious, although probably is still worth attempting.[27]

Obviously, there is nothing one can do about the bias in the *JAS* sample. The *JAS* itself does alleviate the problem by segregating the data by states or finer subdivisions, but it is clearly still present. If the *JAS* were to publish the number of observations in the sample for each area, type of well, and depth class, some evaluation of the problem, at least, could be made.

B) How well does the organization of the sample data by depth class totals or averages represent the sample data themselves?

It is difficult to overemphasize the importance of this question which only the compilers of the *JAS* are in a position to answer. The sample data are aggregated or averaged by depth class at a relatively early stage in their processing, and all remaining procedures are dependent on such aggregates or averages. If the variance within each depth range is large, an enormous and appalling amount of information and efficiency is lost in dealing with the data in this way.[28]

For our purposes, the question is of importance in evaluating the power of our model; and for *any* purposes it is of importance in evaluating the usefulness of the freehand curve-fitting procedure used by the *JAS*. As has been indicated, and as will appear in the results, our model does extremely well in fitting the depth-class means of both the *JAS* sample for 1959 and the *JAS* figures as published for all years. However, if the scatter of actual well data around the curve which passes through, or close to, the depth-class means is large, the power of our model may be substantially less than

---

[25] *Permian Basin Proceeding,* pp. 3987–94; 4050 ff.; 4380 ff.

[26] *Ibid.,* pp. 3997–4000; 4102 ff.; and *JAS*—1959, p. 13 (the *JAS* seems to think this is desirable). The sample also overrepresents wells drilled by majors as against those drilled by independents (*Permian Basin Proceeding,* pp. 3988 and 4031)—a fact which is, of course, not independent of the overrepresentation of deep wells. It is hard to know what results this latter fact has for our purposes, although it is possible that major company drilling costs at a given depth are less than those of independent operators.

[27] Since the *JAS* figures for the population are based on curves drawn through the sample data, this problem is not removed by the blowup to the population.

[28] This is occasionally less of a problem than it seems. Where, as in a good many cases in the 1959 *JAS* sample, only one well in a depth class appears in the sample, the depth class aggregate for the sample is obviously identical with the sample data. Of course, the question of representativeness of the sample is very serious in such cases.

indicated and the results less reliable. A similar remark obviously holds for *any* judgment as to the regularity of the cost-depth relationship which is made from the *JAS* figures as published or from the *JAS* sample. It would aid if the *JAS* would at least publish estimates of variance of the sample points around their freehand curve. As matters now stand, it is impossible to judge how regular the depth-cost relation really is by using the *JAS* figures or any analysis based on them.

C) How valid and reliable are the procedures used in blowing up the sample to secure estimates of population figures?

There are clearly a number of aspects of the *JAS* procedure which are less than ideal. One of these has already been mentioned—the aggregation of the sample data by depth classes before the curve fitting is performed. One result of this is a highly inefficient method of curve fitting. In addition, it would seem that no correction is made, since the curve is apparently fitted to the depth-class aggregates without weighting the observations by the number of wells reporting in each depth class, a procedure that would take account of the fact that there is more information on some depth classes than on others.[29]

The loss of information involved here is serious, the more so because it makes it difficult to perform any more sophisticated curve-fitting operation than freehand drawing. While it would be unrealistic to expect the *JAS* to go through the elaborate non-linear procedures described in Appendix A, it would not be unreasonable to expect a polynomial approximation to (9) to be more reliable than a freehand curve, and this could be easily fitted to the sample data since the number of reporting wells in most cases is fairly large. (This would also facilitate computation of error variances.) The use of depth-class aggregates reduces the available number of degrees of freedom to the point where this is impractical.[30]

Fitting a freehand curve has a further disadvantage in that the curve itself cannot practicably be reported. Since the sample data are not published, there is no alternative to accepting the *JAS* estimates, even though no measure of their reliability is published.

Two minor points are worth mentioning. First, if a freehand curve is to be used, it would be more suitable to fit it to figures on cost per well rather than those on cost per foot, the basic relation being one of cost per well to depth. Since, as will be seen in Section 5, equation (9) works so well, there seems little reason to suppose that division by depth simplifies the function

---

[29] See the preceding footnote. The issue here turns on whether the sample is evenly distributed over depth classes, not on the absolute number of wells reporting in each depth class. A highly uneven distribution in the sample is by no means uncommon, however.

[30] It only requires three observations to fit the two-parameter function (9), but the computational procedure involved is nearly prohibitive (see Appendix A); a polynomial approximation would involve several more parameters and would thus require a good many more observations.

to be fitted. It is realized, of course, that, as individual well depths are not available for the unsampled wells, using cost-per-well figures would make the blowup to the drilling population somewhat more difficult, but a tolerable approximation could be arrived at—and one not more arbitrary than is now involved in some of the procedures now used—by assuming all unreported wells in a depth class to have the same footage.

A second questionable matter is the addition of estimated figures for unreported wells to those for wells in the sample. While it is possible that greater accuracy is attained thereby, it can also be argued that the *JAS* freehand curve itself gives a better set of estimates for all wells taken together. At the least, the constructed figures for unreported wells and the sample figures should be reported separately.

In general, so far as *a priori* argument goes the *JAS* procedure is not likely to be a good one. If, however, the *JAS* were to publish at least the sample depth-class aggregates, the procedure could be systematically evaluated.

In part, such an evaluation can, in fact, be made, though it is limited to data for 1959 made available as a result of the *Permian Basin Proceeding*. The *JAS* sample for this year, organized by depth-class aggregates, does not permit evaluation of answers to questions (*A*) and (*B*), discussed above, but it does cast light on the effects of the *JAS* procedures used to obtain population estimates. Further, it provides for the present analysis data that are actually reported rather than figures that are constructed by previous curve-fitting. The *JAS* sample for 1959 thus constitutes the primary collection of data in this study. It will be used as a basis for comparing the sample with the *JAS* figures as published for that year and hence can give us a measure of the effects of the *JAS* curve-fitting procedures. To some extent, this will enable us to evaluate the results of performing an analysis on the *JAS* figures for 1955 and 1956 and the changes over time that appear to have taken place.

To compare the 1959 sample and the 1959 published figures, equation (9) will be fitted to both sets of data where the number of depth classes reported allows. This will first allow us to see whether the appearance of regularity in the cost-depth relationship is principally a reflection of the *JAS* curve-fitting procedures or whether such regularity is largely present in the sample also.[31] Second, biases in the *JAS* procedure, if any, will show up as systematic differences in the parameters of equation (9) as estimated for the two sets of data. Because these may imply biases of small size in the estimation of marginal or total cost figures, the implied estimates of these

[31] Recall, however, that even the sample figures available are in depth-class means or aggregates. No test of the regularity of the relationship in the original sample can therefore be performed.

After this study was completed, data for individual wells for producers concerned in the *Permian Basin Proceeding* became available (letter of Bruce C. Netschert to the author). It would be interesting to apply the techniques of the present analysis to these data.

magnitudes will be compared at several depths. As one result of this analysis, we shall be able to see whether any or all of these problems are more or less serious in states with a significant amount of drilling, as opposed to others. The entire procedure depends on the use of equation (9), a relationship which provides an excellent fit when it is applied to practically any of the available data.

D) How can the JAS be improved?

Many of the ways in which the *JAS* might be improved have been mentioned in connection with our examination of questions *(A)*, *(B)*, and *(C)*. Here, these suggestions are drawn together with others that seem to warrant attention. The list is doubless not exhaustive.

Desirable as it may be to improve the *JAS* sampling procedure itself, it is probably premature to expect the *JAS* to strive for a stratified sample which correctly represents the depth-formation relation. Because the number of wells reporting in the sample is not a very high percentage of the wells drilled in most areas, it may be more realistic to hope for a serious effort toward fuller coverage.

On the other hand, the use made of the sample as collected can certainly be improved. If it is desirable to obtain estimates for the entire population of wells drilled, it is equally desirable to separate such estimates from those which come directly from actual reported data and to adopt a procedure which can be evaluated. The obvious way to do this is to report the sample data and to forego freehand curve fitting.

If individual well data cannot be reported for reasons of disclosure,[32] the sample depth-class aggregates can be reported. Less satisfactory, but still of great help, would be a system by which the *JAS* would publish the number of wells reporting in each depth class for each type of well and data area, and would compute some variance measure.

Finally, the *JAS* needlessly throws away a great deal of the sample information by using depth-class aggregates in its estimating procedure. There seems to be no good reason why the *JAS* should not use individual well data in making its own estimates, whether or not the data themselves can be published.

The fact that the *JAS* represents one of the best and most complete data-collection jobs in the industry leads one to expect that its data should be fully usable for analysis. This they are not. When the inferences to be drawn from the 1959 sample are compared with the 1959 published figures (as is done in Section 6) serious problems show up in some areas. They are

---

[32] The *JAS* clearly regards such reasons as controlling here, and no question would arise were it not for the fact that in a few cases figures are reported for only one well of a given type in a given depth class. These are included in the sample, so that the depth-class aggregates and the actual individual well figures are identical. Consistency would demand that such cases not be reported.

problems that could be overcome if the *JAS* were to adjust some of its procedures.

5. The Results for the 1959 JAS Sample

The estimating procedures used and the measures of goodness of fit employed are discussed at some length in Appendix A. The numerical results are given in detail in Appendix B, and should be consulted for information on specific areas and for more details than it is possible to convey in the summary tables of this and subsequent sections.

We begin with the results when equation (9),

$$Y = K(e^{\alpha X} - 1),$$

is fitted to the *JAS* 1959 sample depth-class means. These and the goodness-of-fit measures associated with them are given in detail in Table B1 of Appendix B. In general, the results are notable for the excellence of the fit. In nearly all cases, well over 90 per cent of the weighted variance is explained, and frequently over 99 per cent. Coefficients are more often than not more than ten times their asymptotic standard errors and the ratio is frequently much higher. Even J^2—the unweighted measure of goodness of fit which is not being maximized (see Appendix A)—is usually quite high. While significance tests are not known for small samples in nonlinear regressions, it is clear that the results are very good and that strong support is lent by these results to equation (9) as the basic relationship between drilling costs and depths.

The good fits obtained clearly show a strong degree of regularity in the depth-class averages to which the function was fitted. The relation of this to the similar impression given by the *JAS* as published will be discussed later (see page 71); here it is sufficient to point out that such regularity may or may not extend to the individual well data themselves; we have no way of determining this matter.

We turn now to the parameter estimates themselves. Tables 1A–1D (pages 67–70) show the joint distribution of the two fundamental parameters of the model, α (the rate of increase of marginal cost with depth) and H (the limit of marginal cost as depth goes to zero) for dry, productive, oil, and gas wells. The paired numbers in each box represent the number of data areas appearing in that box and the number of wells of the given type drilled in those areas in 1959. These totals do *not* include data areas such as "Total Texas" which are aggregates of other included areas.[33] The distribution of α alone can be read off the row totals in the right-hand column and that of H can be read off the column totals at the bottom. Note that the

---

[33] This system will be used in all later tables.

tables can be read as a rough diagram showing the relation between α and H over all areas.

The most striking thing about these tables is the strong negative association between α and H. Areas with relatively high values of α tend to be areas with relatively low values of H and areas with relatively high values of H tend to be areas with relatively low values of α. While there are a few areas in which both parameters have relatively low values, there are practically none in which both have relatively high values. This is a highly reasonable result. Regions (within data areas) with relatively high values of H are regions with relatively high costs in shallow wells; regions with relatively high values of α are regions with a relatively high rate of increase of cost with depth, given H. It follows that regions with relatively high values of both parameters are regions which are relatively high-cost ones at all depths and we should not expect to see much drilling in them. Similarly, regions in which both parameters have relatively low values are relatively cheap ones and are likely to have been drilled up in the past. We should therefore expect the negative relation observed, other things being equal.[34]

An exception to this argument is the case of a region which for one reason or another has not been explored in the past and can be expected to contain relatively large amounts of petroleum. The presence of large prizes can obviously overcome the cost effects just discussed. Drilling in offshore areas is a clear example of this situation and, accordingly, the estimates for these areas shown in the tables tend to lie slightly below and to the right of the general run of observations, thus indicating a slight, but only slight, tendency for such areas to have a relatively high value of α for their H and a relatively high value of H for their α and hence to be relatively high-cost areas.[35]

What of the order of magnitude of the observed parameter values themselves? In very rough terms, the geographical areas tend to fall into two principal groups: a rather sizeable one including Texas and Louisiana in which α is between 1.5×10^{-4} and 2.5×10^{-4} and H between \$4.00 and \$10.00 per foot; and a somewhat smaller one including California and New Mexico in which α is about 0.5×10^{-4} to 1.0×10^{-4} and H about \$8.00 to \$14.00 per foot. There is a tendency for dry wells to show somewhat higher values of α and lower values of H than do productive, oil, or gas wells, and this seems reasonable, since productive wells have cost items which go up fairly linearly with depth (casing, for example). These items add to marginal costs (and hence to H) a more or less constant amount and thus reduce curvature (measured by α).

Concentrating on the meaning of the order of magnitude of values of α

---

[34] Note that this argument applies chiefly to regions within a given data area, not necessarily to that data area itself as a whole.

[35] Set-up costs are not included.

observed, a better idea of the implications may be obtained if it is noted that a value of 2.0 × 10⁻⁴ corresponds to an increase in marginal cost of about 22 per cent every thousand feet, while one of 1.0 × 10⁻⁴ corresponds to a similar increase of about 10.5 per cent. The two principal groups of areas just described therefore differ rather widely in the rate of increase of marginal cost with depth. Furthermore, the rate of increase of marginal cost with depth is thus quite substantial for a large number of areas; as total cost goes up even faster than marginal cost, the effects of depth are evidently large ones which differ over areas in no trivial way.

As remarked above, another way to look at values of α is in terms of the depth increase in which cost doubles, given the initial depth. Such depth increases are given for selected initial depths for each data area and type of well in Table B2 of Appendix B. A fair idea of what is involved may be obtained from Figure 2 (page 50), which plots the required depth increase as a function of initial depth for selected values of α. Again very roughly, for a value of α of 2.0 × 10⁻⁴, total costs at 7,500 feet are double those at 5,000 feet; total costs at 13,100 feet are double those at 10,000 feet; total costs at 18,400 feet are double those at 15,000 feet; and marginal costs always double in 3,500 feet. These figures are perhaps conservative as representative of the larger of the two principal groups of data areas. For the other group, typical figures corresponding to a value of α of about 0.9 × 10⁻⁴ are as follows: total costs at 8,400 feet are double those at 5,000 feet; total costs at 15,200 feet are double those at 10,000 feet; total costs at 15,000 feet would be doubled at 21,100 feet if any such wells were drilled; and marginal costs always double in 7,700 feet. Clearly, this is a percentage rate of increase of costs with depth that is far slower than that observed in the first of the two groups. Once more, it must be observed that the above figures are very rough and that those for individual areas may differ widely.

Before leaving the results for the *JAS* 1959 sample, one phenomenon emerges from them which is worth remarking. As occurs also in the results reported in Section 6, the parameter estimates for aggregate data areas such as "Total Texas" frequently do *not* lie in the midst of the corresponding estimates for the data areas of which the aggregates are composed. That this can happen without computational error may be seen from the following immensely oversimplified example, demonstrated in Figure 3, in which the estimated functions have been made linear for purposes of expositional clarity.[36]

In Figure 3 the observations for two data areas are respectively indicated by crosses and circles. The estimated relation for the first area is then the line denoted by R_1; that for the second is the line denoted by R_2. As it happens, the first area contains all the high-depth observations and the second one all the low-depth observations. The result when these are aggregated

---

[36] For a much fuller and rigorous discussion of this type of effect, see H. Theil, *Linear Aggregation of Economic Relations* (Amsterdam: North-Holland, 1954).

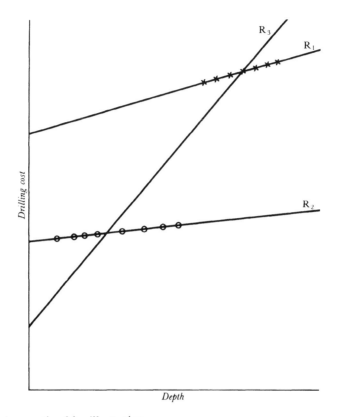

Figure 3. Aggregation bias illustration.

is that the estimated relation is the line denoted by R_3, whose slope is far greater and intercept far less than the corresponding parameters of either R_1 or R_2.

Of course, the example is an extreme one, but it should serve to emphasize the point, already made, which emerges from the presence of this type of phenomenon in our results. What is being estimated for a large data area is largely a depth-formation relationship (this is just what R_3 is in the example) as depth is not distributed independently of type of formation. This is not a consequence of our model; it is endemic in the organization of the *JAS* sample. Such a relationship may be highly interesting, but the fact that the effect noted shows up so clearly when aggregate data areas are used should drive home the danger of drawing conclusions from data for such aggregate areas and, *a fortiori*, from data for the United States as a whole.

TABLE 1A – JAS SAMPLE – 1959; DISTRIBUTION OF α AND H: DRY WELLS

H (current dollars per foot) / α (×10⁻⁴ omitted)	0 - 2	2 - 4	4 - 6	6 - 8	8 - 10	10 - 12	12 - 14	14 - 16	Greater than 16	Totals
0 - .5	0	Indiana Nebraska (2, 1182)	0	Michigan (1, 308)	0	0	0	0	Alaska (1, 9)	(4, 1499)
.5 - 1.0	0	0	Alabama (1, 25)	0	Onshore Calif., Total Calif., Panhandle Tex. (2, 731)	S. E. New Mexico, Total New Mexico (1, 302)	N. W. New Mexico (1, 112)	0	0	(5, 1170)
1.0 - 1.5	0	Illinois (1, 1062)	0	0	0	Utah (1, 119)	0	0	0	(2, 1181)
1.5 - 2.0	0	Mississippi (1, 362)	Arkansas (1, 334)	0	0	Offshore Louisiana (1, 118)	0	0	0	(3, 814)
2.0 - 2.5	0	South La., Onshr. La., N. Dak., Okla., West Tex., Tot. Tex. (4, 4152)	Total Louisiana 0	0	0	0	0	0	0	(4, 4152)
2.5 - 3.0	0	Ky., N. La., Mont., E. Tex., N.Cen. Tex., S.W. Tex., Wyo. (7, 6277)	Appalachian (1, 492)	0	0	0	0	0	0	(8, 6769)
3.0 - 3.5	Gulf Coast Texas (1, 902)	0	0	0	0	0	0	0	0	(1, 902)
3.5 - 4.0	Colorado (1, 560)	0	0	0	0	0	0	0	0	(1, 560)
4.0 - 4.5	Kansas (1, 1937)	0	0	0	0	0	0	0	0	(1, 1937)
TOTALS	(3, 3399)	(15, 13035)	(3, 851)	(1, 308)	(2, 731)	(3, 539)	(1, 112)	0	(1, 9)	(29, 18984)

67

TABLE 1B – JAS SAMPLE – 1959; DISTRIBUTION OF α AND H: PRODUCTIVE WELLS

H (current dollars per foot) / α (× 10⁻⁴ omitted)	0 - 2	2 - 4	4 - 6	6 - 8	8 - 10	10 - 12	12 - 14	14 - 16	Greater than 16	Totals
0 - .5	0	0	0	0	0	Indiana, Montana (2, 485)	0	0	Alaska (1, 9)	(3, 494)
.5 - 1.0	0	0	0	Kentucky (1, 2435)	0	S. E. New Mexico, Total New Mexico (1, 879)	Onshore California (1, 931)	Total California 0	0	(3, 4245)
1.0 - 1.5	0	0	0	0	Panhandle Texas, West Texas (2, 5202)		0	0	0	(2, 5202)
1.5 - 2.0	0	0	South Louisiana, Total Texas, (1, 1137)	Appalachian, Onshore Louisiana, Southwest Texas (2, 3337)	N. W. New Mexico, Wyoming (2, 1235)	Offshore Louisiana, Total Louisiana (1, 323)	0	0	0	(6, 6032)
2.0 - 2.5	0	0	North Dakota, Oklahoma, N. Cen. Texas (3, 6751)	Illinois (1, 1018)	Arkansas, Michigan, Utah (3, 997)		0	0	0	(7, 8766)
2.5 - 3.0	0	Mississippi, East Texas, Gulf Coast Texas (3, 2189)	Colorado (1, 248)	0	0	0	Offshore California (1, 27)	0	0	(5, 2464)
3.0 - 3.5	0	0	0	0	0	0	0	0	0	0
3.5 - 4.0	0	North Louisiana (1, 916)	0	0	0	0	0	0	0	(1, 916)
4.0 - 4.5	0	0	0	0	0	0	0	0	0	0
4.5 - 5.0	0	Kansas (1, 1943)	0	0	0	0	0	0	0	(1, 1943)
TOTALS	0	(5, 5048)	(5, 8136)	(4, 6790)	(7, 7434)	(4, 1687)	(1, 931)	(1, 27)	(1, 9)	(28, 30062)

TABLE 1C – JAS SAMPLE – 1959; DISTRIBUTION OF α AND H: OIL WELLS

α (×10⁻⁴ omitted) \ H (current dollars per foot)	0 - 2	2 - 4	4 - 6	6 - 8	8 - 10	10 - 12	12 - 14	14 - 16	TOTALS
0 - .5	0	0	0	0	Kentucky (1, 2146)	Indiana (1, 295)	Northwest New Mexico, Wyoming (2, 781)	0	(4, 3222)
.5 - 1.0	0	0	0	0	S. E. New Mexico, Total New Mexico (1, 829)	Montana (1, 168)	Onshore California, Total California (1, 857)	0	(3, 1854)
1.0 - 1.5	0	0	0	Panhandle Texas (1, 762)	West Texas (1, 4035)	0	Total Louisiana 0	0	(2, 4797)
1.5 - 2.0	0	0	North Louisiana (1, 752)	Colo., Illinois, S. La., Onshr. La., Tot. Tex. (3, 1987)	0	0	Offshore Louisiana (1, 249)	0	(5, 2988)
2.0 - 2.5	0	0	Kansas, North Dakota, Oklahoma, N. Cen. Texas (4, 7803)	0	Michigan, Utah (2, 424)	0	0	0	(6, 8227)
2.5 - 3.0	0	Gulf Coast Texas (1, 681)	Arkansas, Southwest Texas (2, 1379)	0	0	0	0	Offshore California (1, 27)	(4, 2087)
3.0 - 3.5	Mississippi (1, 251)	0	0	0	0	0	0	0	(1, 251)
3.5 - 4.0	0	East Texas (1, 715)	0	0	0	0	0	0	(1, 715)
4.0 - 4.5	0	0	0	0	0	0	0	0	0
4.5 - 5.0	0	Appalachian (1, 881)	0	0	0	0	0	0	(1, 881)
TOTALS	(1, 251)	(3, 2277)	(7, 9934)	(4, 2749)	(5, 7434)	(2, 463)	(4, 1887)	(1, 27)	(27, 25022,)

TABLE 1D – JAS SAMPLE – 1959; DISTRIBUTION OF α AND H: GAS WELLS

α (×10⁴ omitted) \\ H (current dollars per foot)	0 - 2	2 - 4	4 - 6	6 - 8	8 - 10	10 - 12	12 - 14	14 - 16	16 - 18	TOTALS
0 - .5	0	0	0	0	0	0	0	0	0	0
.5 - 1.0	0	0	0	0	0	0	Onshore California, Total California (1, 74)	0	0	(1, 74)
1.0 - 1.5	0	0	Onshore Louisiana 0	Kentucky, N. Cen. Texas (2, 494)	Panhandle Texas (1, 277)	0	Wyoming (1, 77)	0	Arkansas (1, 41)	(5, 889)
1.5 - 2.0	0	0	0	Appalachian, Mississippi (2, 1214)	Tot. La., S.E. New Mex., Tot. N. Mex., S.W. Tex. W. Tex. (3, 535)	Colorado (1, 88)	0	0	Offshore Louisiana (1, 74)	(7, 1911)
2.0 - 2.5	0	0	Kan., S. La., Glf. Cst. Tex., Tot. Tex. (3, 914)	Oklahoma (1, 498)	0	0	0	0	0	(4, 1412)
2.5 - 3.0	0	East Texas (1, 98)	0	Northwest New Mexico (1, 377)	0	0	0	0	0	(2, 475)
3.0 - 3.5	0	0	0	0	0	0	0	0	0	0
3.5 - 4.0	0	0	0	0	0	0	0	0	0	0
4.0 - 4.5	0	0	0	0	0	0	0	0	0	0
4.5 - 5.0	North Louisiana (1, 164)	0	0	0	0	0	0	0	0	(1, 164)
TOTALS	(1, 164)	(1, 98)	(3, 914)	(6, 2583)	(4, 812)	(1, 88)	(2, 151)	0	(2, 115)	(20, 4925)

6. The JAS Data as Published versus the JAS Sample, 1959

The estimates for the *JAS* data as published for 1955, 1956, and 1959 are given in detail in Appendix B, Table B3. We begin our analysis by comparing the estimates for the data as published for 1959 with those for the sample already discussed. While all the comparisons depend, explicitly or implicitly, on our estimates of equation (9), the estimated curves fit the data from the sample and the data as published so well in nearly all cases that the results of the comparisons can be used in evaluating the *JAS* as published.

Most of the comparisons dealt with here will be made again in Section 7, where changes over time are examined. In the first place, it is obvious that the freehand curve-fitting procedure of the *JAS* must tend to increase the appearance of regularity in the data. The good fits described in Section 5 do indicate that the sample depth-class means are already quite regular; nevertheless, it is worth asking how much smoothing is involved in the *JAS* procedure. One way to arrive at the answer is to take the ratio of the weighted squared coefficients of determination, corrected for degrees of freedom $(\bar{I}^2)$. This measures the per cent increase in regularity involved in moving from the sample to the published figures. However, this may not be an appropriate measure when taken alone. Because $\bar{I}^2$ has an upper limit of unity and most of our estimates from the sample are close to this limit, percentage increases in regularity so measured cannot help but be small. It may be true enough that this circumstance merely reflects the unimportance of the *JAS* data-smoothing procedure in a situation where there is little smoothing to be done. Nevertheless, it would seem appropriate to combine with $\bar{I}^2$ a measure of the percentage reduction in *irregularity*. One might thus plausibly ask how much of the variance of the data around the estimated curve is eliminated in the smoothing procedure. Accordingly, we take the ratio of the standard errors of estimate (S) as a measure of this.[37]

These comparisons, as well as those later discussed, are given in detail in Appendix B, Table B4. As expected, the *JAS* procedure tends to increase the appearance of regularity in the figures. While percentage changes in $\bar{I}^2$ are usually small, they are generally positive, and the percentage reduction in unexplained variance is frequently quite large. It is worth reiterating that this comparison only refers to the effects of the *JAS* freehand curve-fitting procedure; it may well be that the organization of the sample into depth-class means by the *JAS* greatly increases the appearance of regularity in our estimates (or anyone else's) from the individual well data themselves.

We turn now to the more important question of the possible introduction of bias in the parameters of equation (9) as estimated and in the cost esti-

---

[37] Actually, the ratio of standard errors should be squared to give a measure fully comparable with the ratio of $\bar{I}^2$'s. The conclusions to be drawn do not depend on this, however, as no exact combination of the two measures is being suggested.

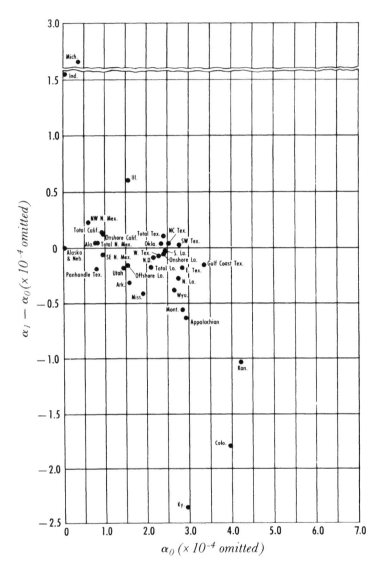

Figure 4A. DRY WELLS. *JAS* data for states as published versus *JAS* sample, 1959: $(\alpha_1 - \alpha_0)$ plotted against α_0. (Source: Appendix Tables B1, B4.)

mates themselves. Since examination of the properties of (9) showed that it is the actual rather than the percentage difference in α which is of crucial importance in differences in cost estimates, we begin with that difference.[38]

Throughout this section, the subscript 1 denotes the estimate from the published data and the subscript 0 denotes that from the sample. The dis-

---

[38] The percentage differences in α may be found in Appendix B, Table B4, as may the percentage differences in K which are not separately discussed in the text.

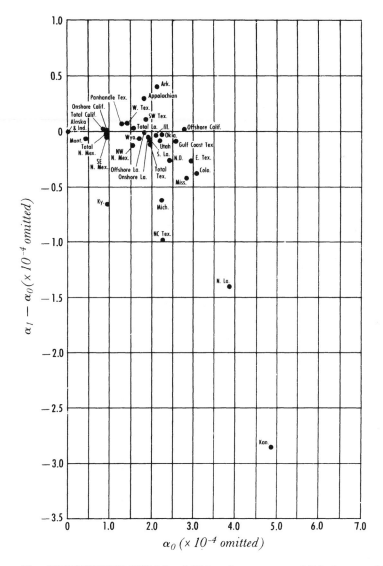

Figure 4B.   PRODUCTIVE WELLS.   *JAS* data for states as published versus *JAS* sample, 1959:
$(\alpha_1 - \alpha_0)$ plotted against α_0. (Source: Appendix Tables B1, B4.)

tribution of $\alpha_1 - \alpha_0$ is given in Tables 2A–2D (pages 77–78) for the different well types.

These tables show that there is a clear tendency for α as estimated from the *JAS* published data to be lower than the corresponding true value as estimated from the *JAS* sample. Moreover, that tendency is greater, the higher is the sample-estimated value of α; this may be seen from Figures 4A–4D in which $\alpha_1 - \alpha_0$ is plotted against α_0 for the different well types.

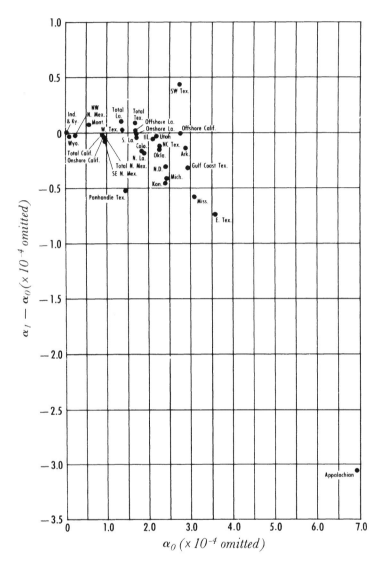

Figure 4C.   OIL WELLS.   *JAS* data for states as published versus *JAS* sample, 1959:
    $(\alpha_1 - \alpha_0)$ plotted against α_0. (Source: Appendix Tables B1, B4.)

This result is quite serious and does not simply relate to estimates of equation (9). It means that there is a pronounced tendency for the *JAS* to understate the curvature of the cost-depth relationship—to bias the appearance of that relationship toward linearity. This is not surprising; it is a natural consequence of freehand curve fitting that the curvature of a strongly curvilinear relationship tends to be understated.

As one should expect, since the *JAS* freehand curves presumably pass close to the sample depth-class means in intermediate depth ranges, the

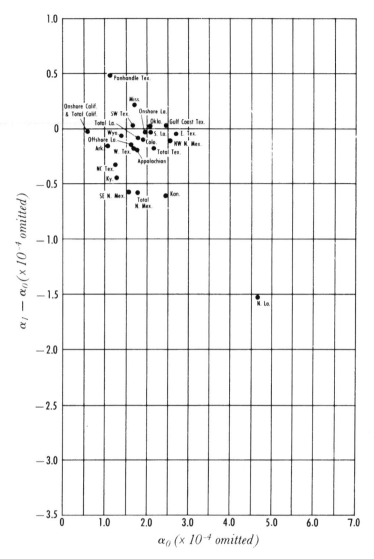

Figure 4D.   GAS WELLS.   *JAS* data for states as published versus *JAS* sample, 1959:
$(\alpha_1 - \alpha_0)$ plotted against α_0. (Source: Appendix Tables B1, B4.)

understatement in α tends to be accompanied in the same data areas by an overstatement in H (and in K) which compensates for it in part. This may be seen from the distribution of H_1/H_0 presented in Tables 3A–3D (pages 79–80). Moreover, since α is the percentage change in marginal cost with depth, the result of an understatement in α and an overstatement in H is an overstatement of marginal cost in shallow ranges and an understatement in deep ranges. Again, this is the predominant, but not universal tendency, the opposite being true in the minority of cases in which α is overstated. It is also occasionally the case that marginal cost is everywhere overstated or understated, the degree of such error varying with depth. Denoting marginal cost by MC, the distributions of MC_1/MC_0 at 5,000, 10,000, and 15,000 feet are given in Tables 4A–4L (pages 81–86).

As one should expect, the effect in question shows up most strongly in the case of α, less strongly in the case of marginal cost, and least strongly in the case of total cost. Nevertheless it is clearly present in the last and most important case also. The distributions of Y_1/Y_0 at 5,000, 10,000, and 15,000 feet are presented in Tables 5A–5L (pages 87–92). Note that the corresponding distributions near the surface are those of H_1/H_0 already presented.[39] While the estimated total costs from the published data frequently come reasonably close to those from the sample, there are some strong exceptions. Even for wells as deep as 5,000 feet (a depth somewhat greater than that of the average well), total costs tend to be overstated, and the depth at which that tendency disappears seems to be somewhere in the neighborhood of 10,000 feet.[40] Costs are thus overstated for the large majority of wells.

To sum up: The *JAS* data as published tend to overstate costs at average well depths and to understate the effects of depth on cost, although there are frequent exceptions to this rule. The latter understatement tends to be greatest where the effects themselves are largest. It seems clear that the *JAS* should make a practice of releasing at least the sample depth-class totals as well as population estimates. As now presented, the published figures can be rather misleading.

We must now consider the implications of the foregoing for comparisons over time of estimates from the *JAS* published figures. Clearly, the presence of bias in the published figures can affect the reliability of the comparisons. On the other hand, to the extent that biases persist over time, they will cancel out of inter-year comparisons. Because the *JAS* for 1955–56 did not use freehand curve drawing, its results may well be less biased than those of the 1959 survey just examined, but, lacking the appropriate data, it is impossible to test this hypothesis. To the extent that it holds, comparisons of 1959 with 1955–56 made exclusively from the published data will be affected largely by the biases found in the 1959 survey. On the other hand, comparison of the 1955–56 published data with the 1959 survey would leave an unknown amount of bias (possibly very little) stemming from the 1955–56 procedures. It seems prudent to use the published data in both cases on the assumption that biases in the 1955–56 survey will also appear in the 1959 one, and to check our results in terms of the comparisons made in this section. Fortunately, when one alters the inter-year comparisons by replacing the 1959 data as published with the 1959 sample, the over-all conclusions presented in Section 7 are not weakened, and in some cases are strengthened.

---

[39] By L'Hôpital's Rule.

[40] A little experimentation will show that an overstatement of costs at low depth and an understatement at high depth is the natural tendency when fitting a freehand curve to points which really lie on a function as concave from above as (9) is, given the orders of magnitude of the parameters as estimated from the sample data.

TABLE 2A – JAS DATA AS PUBLISHED versus JAS SAMPLE – 1959
Distribution of $\alpha_1 - \alpha_0$ ($\times 10^{-4}$ omitted): DRY WELLS

-2.5 to -2.25	-2.25 to -2.0	-2.0 to -1.75	-1.75 to -1.5	-1.5 to -1.25	-1.25 to -1.0	-1.0 to -.75	-.75 to -.5	-.5 to -.25	.25 to 0-
Kentucky		Colorado		Kansas			Appalachian Montana	Arkansas, North Louisiana, Mississippi, Wyoming	South Louisiana, Onshore Louisiana, Offshore Louisiana, Southeast New Mexico, North Dakota / East Texas, Gulf Coast Texas, Panhandle Texas, West Texas, Utah
(1, 893)	0	(1, 560)	0	0	(1, 1937)	0	(2; 661)	(4, 1798)	(9, 4171)

0^a	0+ to .25	.25 to .5	.5 to .75	.75 to 1.0	1.0 to 1.25	1.25 to 1.5	1.5 to 1.75	1.75 to 2.0	2.0 to 2.25	2.25 to 2.5	2.5 to 2.75
Alaska, Nebraska	Alabama, Onshore California, Total California, Total Louisiana / Northwest New Mexico, Total New Mexico, North Central Texas, Southeast Texas, Total Texas		Illinois				Indiana[b]				Michigan
(2, 625)	(6, 6403)	0	(1, 1062)	0	0	0	(1, 566)	0	0	0	(1, 308)

Total less than 0	Total 0^a	Total greater than 0	GRAND TOTAL
(18, 10020)	(2, 625)	(9, 8339)	(29, 18984)

[a]Both estimates are linear and unreliable. (See Appendix A.)

[b]Sample estimate is linear and unreliable. (See Appendix A.)

TABLE 2B – JAS DATA AS PUBLISHED versus JAS SAMPLE – 1959
Distribution of $\alpha_1 - \alpha_0$ ($\times 10^{-4}$ omitted): PRODUCTIVE WELLS

-3.25 to -3.0	-3.0 to -2.75	-2.75 to -2.5	-2.5 to -2.25	-2.25 to -2.0	-2.0 to -1.75	-1.75 to -1.5	-1.5 to -1.25	1.25 to -1.0	-1.0 to -.75	-.75 to -.5	-.5 to -.25
	Kansas						North Louisiana		North Central Texas	Kentucky, Michigan	Colorado, Mississippi, North Dakota, East Texas
0	(1, 1943)	0	0	0	0	0	(1, 916)	0	(1, 3298)	(2, 2730)	(4, 1622)

-.25 to 0-	0^a	0+ to .25	.25 to .5
Illinois, South Louisiana, Onshore Louisiana, Offshore Louisiana, Montana, Northwest New Mexico / Southeast New Mexico, Total New Mexico, Oklahoma, Total Texas, Utah, Wyoming	Alaska, Indiana	Onshore California, Offshore California, Total California, Total Louisiana / Gulf Coast Texas, Panhandle Texas, Southwest Texas, West Texas	Appalachian, Arkansas
(9, 8145)	(2, 315)	(6, 8527)	(2, 2566)

Total less than 0	Total 0^a	Total greater than 0	GRAND TOTAL
(18, 18654)	(2, 315)	(8, 11093)	(28, 30062)

[a]Both estimates are linear and unreliable. (See Appendix A.)

77

TABLE 2C – JAS DATA AS PUBLISHED versus JAS SAMPLE – 1959
Distribution of $a_1 - a_0$ ($\times 10^{-4}$ omitted): OIL WELLS

-3.25 to -3.0	-3.0 to -2.75	-2.75 to -2.5	-2.5 to -2.25	-2.25 to -2.0	-2.0 to -1.75	-1.75 to -1.5	-1.5 to -1.25	-1.25 to -1.0	-1.0 to -.75	-.75 to -.5	-.5 to -.25
Appalachian									East Texas	Mississippi Panhandle Texas	Kansas Michigan North Dakota Gulf Coast Texas
(1, 881)	0	0	0	0	0	0	0	0	(1, 715)	(2, 1013)	(4, 2955)

-.25 to 0-		0ᵃ	0+ to .25	.25 to .5
Arkansas Onshore California Total California Colorado Illinois North Louisiana South Louisiana Onshore Louisiana	Northwest New Mexico Southeast New Mexico Total New Mexico Oklahoma North Central Texas Utah Wyoming	Indiana Kentucky	Offshore California Offshore Louisiana Total Louisiana Montana West Texas Total Texas	Southwest Texas
	(12, 11621)	(2, 2441)	(4, 4479)	(1, 917)

Total less than 0	Total 0ᵃ	Total greater than 0	GRAND TOTAL
(20, 17185)	(2, 2441)	(5, 5396)	(27, 25022)

ᵃBoth estimates are linear and unreliable. (See Appendix A.)

TABLE 2D – JAS DATA AS PUBLISHED versus JAS SAMPLE – 1959
Distribution of $a_1 - a_0$ ($\times 10^{-4}$ omitted): GAS WELLS

-1.75 to -1.5	-1.5 to -1.25	-1.25 to -1.0	-1.0 to -.75	-.75 to -.5	-.5 to -.25
North Louisiana				Kansas Southeast New Mexico Total New Mexico	Kentucky North Central Texas
(1, 164)	0	0	0	(2, 233)	(2, 494)

-.25 to 0-		0	0+ to .25	.25 to .5
Appalachian Arkansas Onshore California Total California Colorado South Louisiana Onshore Louisiana	Offshore Louisiana Total Louisiana Northwest New Mexico East Texas West Texas Total Texas Wyoming	0	Mississippi Oklahoma Gulf Coast Texas Southwest Texas	Panhandle Texas
	(10, 2458)	0	(4, 1299)	(1, 277)

Total less than 0	Total 0	Total greater than 0	GRAND TOTAL
(15, 3349)	0	(5, 1576)	(20, 4925)

TABLE 3A — JAS DATA AS PUBLISHED versus JAS SAMPLE — 1959
Distribution of H_1/H_0: DRY WELLS

.5 to .6	.6 to .7	.7 to .8	.8 to .9	.9 to 1.0-	1.0	1.0+ to 1.1	1.1 to 1.2	1.2 to 1.3
Michigan			Onshore California Total California Indiana Total Louisiana Northwest New Mexico	Alabama Illinois Total New Mexico North Central Texas Southwest Texas Total Texas		Alaska South Louisiana Onshore Louisiana Southeast New Mexico North Dakota West Texas	Arkansas North Louisiana Offshore Louisiana Nebraska — East Texas Gulf Coast Texas Panhandle Texas Utah	
(1, 308)	0	0	(3, 1188)	(5, 6843)	0	(5, 2391)	(8, 3340)	0

1.3 to 1.4	1.4 to 1.5	1.5 to 1.6	1.6 to 1.7	1.7 to 1.8	1.8 to 1.9	1.9 to 2.0	2.0 to 2.1	2.1 to 2.2	2.2 to 2.3	2.3 to 2.4
Kentucky Mississippi	Appalachian Kansas Wyoming	Montana								Colorado
(2, 1255)	(3, 2930)	(1, 169)	0	0	0	0	0	0	0	(1, 560)

Total less than 1	Total 1	Total greater than 1	GRAND TOTAL
(9, 8339)	0	(20, 10645)	(29, 18984)

TABLE 3B — JAS DATA AS PUBLISHED versus JAS SAMPLE — 1959
Distribution of H_1/H_0: PRODUCTIVE WELLS

.7 to .8	.8 to .9	.9 to 1.0-	1.0	1.0+ to 1.1	1.1 to 1.2	1.2 to 1.3	1.3 to 1.4
Arkansas	Total Louisiana	Appalachian Offshore California Total California Gulf Coast Texas Southwest Texas West Texas		Alaska Onshore California Illinois Indiana South Louisiana Onshore Louisiana Offshore Louisiana Michigan — Montana Northwest New Mexico Southeast New Mexico Total New Mexico Oklahoma Panhandle Texas Total Texas Utah	Kentucky Wyoming	Colorado North Dakota East Texas	North Central Texas
(1, 503)	0	(5, 8620)	0	(13, 10277)	(2, 2883)	(3, 1339)	(1, 3298)

1.4 to 1.5	1.5 to 1.6	1.6 to 1.7	1.7 to 1.8	1.8 to 1.9	1.9 to 2.0	2.0 to 2.1	2.1 to 2.2	2.2 to 2.3	2.3 to 2.4	2.4 to 2.5	2.5 to 2.6	2.6 to 2.7
	Mississippi								North Louisiana			Kansas
0	(1, 283)	0	0	0	0	0	0	0	(1, 916)	0	0	(1, 1943)

Total less than 1	Total 1	Total greater than 1	GRAND TOTAL
(6, 9123)	0	(22, 20939)	(28, 30062)

TABLE 3C – JAS DATA AS PUBLISHED versus JAS SAMPLE – 1959
Distribution of H_1/H_0: OIL WELLS

.8 to .9	.9 to 1.0–	1.0	1.0+ to 1.0	
Total Louisiana Southwest Texas	Arkansas Offshore California Offshore Louisiana Montana West Texas Total Texas		Onshore California Total California Colorado Illinois Indiana Kentucky North Louisiana South Louisiana	Onshore Louisiana Northwest New Mexico Southeast New Mexico Total New Mexico Oklahoma North Central Texas Utah Wyoming
(1, 917)	(5, 4941)	0	(13, 13600)	

1.1 to 1.2	1.2 to 1.3	1.3 to 1.4	1.4 to 1.5	1.5 to 1.6	1.6 to 1.7	1.7 to 1.8
Michigan	Kansas North Dakota Gulf Coast Texas Panhandle Texas				Appalachian East Texas	Mississippi
(1, 241)	(4, 3476)	0	0	0	(2, 1596)	(1, 251)

Total less than 1	Total 1	Total greater than 1	GRAND TOTAL
(6, 5858)	0	(21, 19164)	(27, 25022)

TABLE 3D – JAS DATA AS PUBLISHED versus JAS SAMPLE – 1959
Distribution of H_1/H_0: GAS WELLS

.8 to .9	.9 to 1.0–	1.0	1.0+ to 1.1	1.1 to 1.2	1.2 to 1.3
Mississippi Panhandle Texas	Onshore California Total California Oklahoma Gulf Coast Texas Southwest Texas		Colorado South Louisiana Onshore Louisiana Total Louisiana Northwest New Mexico East Texas Total Texas Wyoming	Appalachian Arkansas Kentucky Offshore Louisiana North Central Texas West Texas	Kansas Southeast New Mexico Total New Mexico
(2, 309)	(4, 1341)	0	(5, 959)	(6, 1919)	(2, 233)

1.3 to 1.4	1.4 to 1.5	1.5 to 1.6	1.6 to 1.7	1.7 to 1.8	1.8 to 1.9	1.9 to 2.0	2.0 to 2.1	2.1 to 2.2	2.2 to 2.3	2.3 to 2.4	2.4 to 2.5	2.5 to 2.6	2.6 to 2.7	2.7 to 2.8	2.8 to 2.9	2.9 to 3.0
																North Louisiana
0	0	0	0	0	0	0	0	0	0	0	0	0	0	0	0	(1, 164)

Total less than 1	Total 1	Total greater than 1	GRAND TOTAL
(6, 1650)	0	(14, 3275)	(20, 4925)

80

TABLE 4A – JAS DATA AS PUBLISHED versus JAS SAMPLE – 1959
Distribution of MC_1/MC_0 at 5,000 feet: DRY WELLS

.4 to .5	.5 to .6	.6 to .7	.7 to .8	.8 to .9	.9 to 1.0–		1.0	1.0+ to 1.1	
Kentucky					Alabama Arkansas Onshore California Total California Colorado Kansas Total Louisiana Northwest New Mexico	Southeast New Mexico Total New Mexico Oklahoma North Central Texas Panhandle Texas Southwest Texas Total Texas		Alaska Appalachian North Louisiana South Louisiana Onshore Louisiana Offshore Louisiana	North Dakota East Texas Gulf Coast Texas West Texas Utah
(1, 893)	0	0	0	0	(11, 9757)		0	(10, 4750)	

1.1 to 1.2	1.2 to 1.3	1.3 to 1.4	1.4 to 1.5	1.5 to 1.6	1.6 to 1.7	1.7 to 1.8	1.8 to 1.9	1.9 to 2.0	2.0 to 2.1	2.1 to 2.2
Mississippi Montana Nebraska Wyoming	Illinois						Indiana			Michigan
(4, 1648)	(1, 1062)	0	0	0	0	0	(1, 566)	0	0	(1, 308)

Total less than 1	Total 1	Total greater than 1	GRAND TOTAL
(12, 10650)	0	(17, 8334)	(29, 18984)

TABLE 4B – JAS DATA AS PUBLISHED versus JAS SAMPLE – 1959
Distribution of MC_1/MC_0 at 10,000 feet: DRY WELLS

.1 to .2	.2 to .3	.3 to .4	.4 to .5	.5 to .6	.6 to .7	.7 to .8	.8 to .9	.9 to 1.0–		1.0
Kentucky		Colorado	Kansas			Appalachian	Arkansas Montana	North Louisiana Offshore Louisiana Mississippi Southeast New Mexico North Dakota Oklahoma	East Texas Gulf Coast Texas Panhandle Texas Southwest Texas Utah Wyoming	Total New Mexico
(1, 893)	0	0	(1, 560)	(1, 1937)	0	(1, 492)	(2, 503)	(12, 6916)		0

1.0+ to 1.1		1.1 to 1.2	1.2 to 1.3	1.3 to 1.4	1.4 to 1.5	1.5 to 1.6	1.6 to 1.7	1.7 to 1.8	3.9 to 4.0	8.2 to 8.3
Alabama Alaska Onshore California Total California South Louisiana Onshore Louisiana	Total Louisiana Northwest New Mexico North Central Texas West Texas Total Texas	Nebraska						Illinois	Indiana	Michigan
(7, 5131)		(1, 616)	0	0	0	0	0	(1, 1062)	(1, 566)	(1, 308)

Total less than 1	Total 1	Total greater than 1	GRAND TOTAL
(18, 11301)	0	(11, 7683)	(29, 18984)

81

0 to .1	.1 to .2	.2 to .3	.3 to .4	.4 to .5	.5 to .6	.6 to .7	.7 to .8	.8 to .9
Kentucky	Colorado		Kansas		Appalachian	Montana	Arkansas North Louisiana Mississippi Wyoming	Panhandle Texas Utah
(1, 893)	(1, 560)	0	(1, 1937)	0	(1, 492)	(1, 169)	(4, 1798)	(2, 340)

.9 to 1.0−		1.0	1.0+ to 1.1		1.1 to 1.2	2.3 to 3.4	8.7 to 8.8	31.4 to 31.5
South Louisiana Onshore Louisiana Offshore Louisiana Southeast New Mexico North Dakota	East Texas Gulf Coast Texas Southwest Texas West Texas		Alabama Alaska Onshore California Total California Total Louisiana	Total New Mexico Oklahoma North Central Texas Total Texas	Nebraska Northwest New Mexico	Illinois	Indiana	Michigan
(8, 4962)		0	(5, 5169)		(2, 728)	(1, 1062)	(1, 566)	(1, 308)

Total less than 1	Total 1	Total greater than 1	GRAND TOTAL
(19, 11151)	0	(10, 7833)	(29, 18984)

.6 to .7	.7 to .8	.8 to .9	.9 to 1.0−	1.0
Kansas	Michigan	Arkansas Kentucky North Central Texas	Total California Illinois Total Louisiana Northwest New Mexico Gulf Coast Texas Utah	Offshore California
(1, 1943)	(1, 295)	(3, 6236)	(4, 3097)	(1, 27)

1.0+ to 1.1			1.1 to 1.2	1.2 to 1.3
Alaska Appalachian Onshore California Colorado Indiana South Louisiana Onshore Louisiana	Offshore Louisiana Montana Southeast New Mexico Total New Mexico North Dakota (16, 17265)	Oklahoma East Texas Panhandle Texas Southwest Texas West Texas Total Texas Wyoming	North Louisiana (1, 916)	Mississippi (1, 283)

Total less than 1	Total 1	Total greater than 1	GRAND TOTAL
(9, 11571)	(1, 27)	(18, 18464)	(28, 30062)

TABLE 4E – JAS DATA AS PUBLISHED versus JAS SAMPLE – 1959
Distribution of MC_1/MC_0 at 10,000 feet: PRODUCTIVE WELLS

.1 to .2	.2 to .3	.3 to .4	.4 to .5	.5 to .6	.6 to .7	.7 to .8	.8 to .9
Kansas				Kentucky North Louisiana Michigan North Central Texas			Colorado
(1, 1943)	0	0	0	(4, 6944)	0	0	(1, 248)

.9 to 1.0–			1.0	1.0+ to 1.0		1.1 to 1.2	1.2 to 1.3
Illinois South Louisiana Onshore Louisiana Total Louisiana Mississippi	Montana Northwest New Mexico Southeast New Mexico Total New Mexico North Dakota	Oklahoma East Texas Gulf Coast Texas Total Texas Utah	Offshore California	Alaska Arkansas Onshore California Total California Indiana Offshore Louisiana	Southwest Texas West Texas Wyoming	Panhandle Texas	Appalachian
(11, 9841)			(1, 27)	(8, 7957)		(1, 1039)	(1, 2063)

Total less than 1	Total 1	Total greater than 1	GRAND TOTAL
(17, 18976)	(1, 27)	(10, 11059)	(28, 30062)

TABLE 4F – JAS DATA AS PUBLISHED versus JAS SAMPLE – 1959
Distribution of MC_1/MC_0 at 15,000 feet: PRODUCTIVE WELLS

0 to .1	.1 to .2	.2 to .3	.3 to .4	.4 to .5	.5 to .6	.6 to .7	.7 to .8	.8 to .9
Kansas		North Louisiana	North Central Texas	Kentucky Michigan		Colorado		Mississippi Northwest New Mexico North Dakota East Texas Total Texas
(1, 1943)	0	(1, 916)	(1, 3298)	(2, 2730)	0	(1, 248)	0	(4, 2161)

.9 to 1.0–		1.0	1.0+ to 1.1		1.1 to 1.2	1.2 to 1.3	1.3 to 1.4	1.4 to 1.5
Illinois South Louisiana Onshore Louisiana Total Louisiana Montana	Southeast New Mexico Total New Mexico Oklahoma Utah		Alaska Onshore California Offshore California Total California Indiana	Offshore Louisiana Gulf Coast Texas West Texas Wyoming	Panhandle Texas Southwest Texas		Arkansas	Appalachian
(6, 6587)		0	(8, 7300)		(2, 2313)	0	(1, 503)	(1, 2063)

Total less than 1	Total 1	Total greater than 1	GRAND TOTAL
(16, 17883)	0	(12, 12179)	(28, 30062)

83

TABLE 4G — JAS DATA AS PUBLISHED versus JAS SAMPLE — 1959
Distribution of MC_1/MC_0 at 5,000 feet: OIL WELLS

.3 to .4	.4 to .5	.5 to .6	.6 to .7	.7 to .8	.8 to .9	.9 to 1.0–	1.0
Appalachian					Total Louisiana	Arkansas Total California Colorado Illinois Kansas North Louisiana Offshore Louisiana Michigan	Offshore California
(1, 881)	0	0	0	0	0	(7, 4633)	(1, 27)

1.0+ to 1.1			1.1 to 1.2	1.2 to 1.3
Onshore California Indiana Kentucky South Louisiana Onshore Louisiana Montana Northwest New Mexico	Southeast New Mexico Total New Mexico North Dakota Oklahoma Gulf Coast Texas	North Central Texas Panhandle Texas Southwest Texas West Texas Total Texas Utah Wyoming	East Texas	Mississippi
	(16, 18515)		(1, 715)	(1, 251)

Total less than 1	Total 1	Total greater than 1	GRAND TOTAL
(8, 5514)	(1, 27)	(18, 18481)	(27, 25022)

TABLE 4H — JAS DATA AS PUBLISHED versus JAS SAMPLE — 1959
Distribution of MC_1/MC_0 at 10,000 feet: OIL WELLS

0 to .1	.1 to .2	.2 to .3	.3 to .4	.4 to .5	.5 to .6	.6 to .7	.7 to .8	.8 to .9	.9 to 1.0–	
Appalachian							Kansas Michigan East Texas Panhandle Texas	Arkansas Colorado Gulf Coast Texas	Onshore California Total California Illinois North Louisiana Total Louisiana Mississippi	Northwest New Mexico Southeast New Mexico North Dakota Oklahoma North Central Texas Utah
(1, 881)	0	0	0	0	0	0	(4, 3478)	(3, 1303)	(10, 10334)	

1.0	1.0+ to 1.1		1.1 to 1.2	1.2 to 1.3	1.3 to 1.4
Offshore California	Indiana Kentucky South Louisiana Onshore Louisiana Offshore Louisiana	Montana Total New Mexico West Texas Total Texas Wyoming			Southwest Texas
(1, 27)	(7, 8082)		0	0	(1, 917)

Total less than 1	Total 1	Total greater than 1	GRAND TOTAL
(18, 15996)	(1, 27)	(8, 8999)	(27, 25022)

TABLE 4I – JAS DATA AS PUBLISHED versus JAS SAMPLE – 1959
Distribution of MC_1/MC_0 at 15,000 feet: OIL WELLS

0 to .1	.1 to .2	.2 to .3	.3 to .4	.4 to .5	.5 to .6	.6 to .7	.7 to .8	.8 to .9
Appalachian					Kansas East Texas	Michigan Panhandle Texas	Arkansas[a] Mississippi North Dakota Gulf Coast Texas	Colorado North Louisiana Oklahoma
(1, 881)	0	0	0	0	(2, 2475)	(2, 1003)	(4, 1667)	(3, 3589)

.9 to 1.0-		1.0	1.0+ to 1.1		1.1 to 1.2	1.2 to 1.3	1.3 to 1.4	1.4 to 1.5	1.5 to 1.6	1.6 to 1.7
Onshore California Total California Illinois South Louisiana Total Louisiana Montana	Northwest New Mexico Southeast New Mexico Total New Mexico North Central Texas Utah		Offshore California Indiana Kentucky Onshore Louisiana	Offshore Louisiana West Texas Total Texas Wyoming						Southwest Texas
(8, 7367)		0	(6, 7123)		0	0	0	0	0	(1, 917)

Total less than 1	Total 1	Total greater than 1	GRAND TOTAL
(20, 16982)	0	(7, 8040)	(27, 25022)

[a] .800 to three decimal places.

TABLE 4J – JAS DATA AS PUBLISHED versus JAS SAMPLE – 1959
Distribution of MC_1/MC_0 at 5,000 feet: GAS WELLS

.8 to .9	.9 to 1.0-		1.0	1.0+ to 1.1	
Kentucky Mississippi[a]	Onshore California Total California Kansas Northwest New Mexico Southeast New Mexico	Total New Mexico Oklahoma Gulf Coast Texas North Central Texas Southwest Texas		Appalachian Arkansas Colorado South Louisiana Onshore Louisiana Offshore Louisiana	Total Louisiana East Texas Panhandle Texas West Texas Wyoming
(2, 321)	(8, 2156)		0	(9, 2284)	

1.1 to 1.2	1.2 to 1.3	1.3 to 1.4
		North Louisiana
0	0	(1, 164)

Total less than 1	Total 1	Total greater than 1	GRAND TOTAL
(10, 2477)	0	(10, 2448)	(20, 4925)

[a] .900 to three decimal places.

.6 to .7	.7 to .8	.8 to .9	.9 to 1.0−	
Kansas North Louisiana Total New Mexico	Kentucky Southeast New Mexico	North Central Texas	Appalachian Arkansas Onshore California Total California South Louisiana Onshore Louisiana	Total Louisiana Northwest New Mexico Oklahoma Gulf Coast Texas West Texas Total Texas
(2, 347)	(2, 339)	(1, 205)	(8, 3031)	

1.0	1.0+ to 1.1		1.1 to 1.2	1.2 to 1.3	1.3 to 1.4
	Colorado Offshore Louisiana Mississippi	East Texas Southwest Texas Wyoming			Panhandle Texas
0	(6, 726)		0	0	(1, 277)

Total less than 1	Total 1	Total greater than 1	GRAND TOTAL
(13, 3922)	0	(7, 1003)	(20, 4925)

TABLE 4L – JAS DATA AS PUBLISHED versus JAS SAMPLE – 1959
Distribution of MC_1/MC_0 at 15,000 feet: GAS WELLS

.2 to .3	.3 to .4	.4 to .5	.5 to .6	.6 to .7	.7 to .8	.8 to .9
North Louisiana	Kansas[a] Kentucky	Southeast New Mexico Total New Mexico	North Central Texas		Appalachian Northwest New Mexico West Texas Total Texas	
(1, 164)	0	(2, 472)	(1, 50)	(1, 205)	0	(3, 1687)

.9 to 1.0−		1.0	1.0+ to 1.1	1.1 to 1.2	1.2 to 1.3	1.3 to 1.4	1.4 to 1.5	1.5 to 1.6	1.6 to 1.7	1.7 to 1.8
Arkansas Onshore California Total California Colorado South Louisiana	Onshore Louisiana Offshore Louisiana Total Louisiana East Texas Wyoming		Oklahoma Gulf Coast Texas Southwest Texas	Mississippi						Panhandle Texas
(7, 771)		0	(3, 1267)	(1, 32)	0	0	0	0	0	(1, 277)

Total less than 1	Total 1	Total greater than 1	GRAND TOTAL
(15, 3349)	0	(5, 1576)	(20, 4925)

[a] .500 to three decimal places.

86

TABLE 5A – JAS DATA AS PUBLISHED versus JAS SAMPLE – 1959
Distribution of Y_1/Y_0 at 5,000 feet: DRY WELLS

.8 to .9	.9 to 1.0 –		1.0	1.0+ to 1.1	
Alabama Total Louisiana Northwest New Mexico	Onshore California Total California Kentucky Total New Mexico	Oklahoma North Central Texas Southwest Texas Total Texas		Alaska Arkansas Kansas South Louisiana Onshore Louisiana Offshore Louisiana	Southeast New Mexico North Dakota Gulf Coast Texas Panhandle Texas West Texas
(2, 137)	(5, 7159)		0	(10, 5903)	

1.1 to 1.2	1.2 to 1.3	1.3 to 1.4
Illinois North Louisiana Nebraska East Texas Utah	Appalachian Indiana Michigan Mississippi Wyoming	Colorado Montana
(5, 2827)	(5, 2229)	(2, 729)

Total less than 1	Total 1	Total greater than 1	GRAND TOTAL
(7, 7296)	0	(22, 11688)	(29, 18984)

TABLE 5B – JAS DATA AS PUBLISHED versus JAS SAMPLE – 1959
Distribution of Y_1/Y_0 at 10,000 feet: DRY WELLS

.6 to .7	.7 to .8	.8 to .9	.9 to 1.0–		1.0	1.0+ to 1.1	
Colorado Kansas Kentucky			Alabama Appalachian Arkansas Onshore California Total California North Louisiana Total Louisiana Northwest New Mexico	Southeast New Mexico Total New Mexico Oklahoma North Central Texas Panhandle Texas Southwest Texas Total Texas		Alaska South Louisiana Onshore Louisiana Offshore Louisiana Mississippi Montana	North Dakota East Texas Gulf Coast Texas West Texas Utah Wyoming
(3, 3390)	0	0	(11, 8353)		0	(11, 4689)	

1.1 to 1.2	1.2 to 1.3	1.3 to 1.4	2.0 to 2.1	3.0 to 3.1
Nebraska		Illinois	Indiana	Michigan
(1, 616)	0	(1, 1062)	(1, 566)	(1, 308)

Total less than 1	Total 1	Total greater than 1	GRAND TOTAL
(14, 11743)	0	(15, 7241)	(29, 18984)

87

TABLE 5C – JAS DATA AS PUBLISHED versus JAS SAMPLE – 1959
Distribution of Y_1/Y_0 at 15,000 feet: DRY WELLS

.2 to .3	.3 to .4	.4 to .5	.5 to .6	.6 to .7	.7 to .8	.8 to .9	.9 to 1.0−	
Colorado	Kentucky	Kansas			Appalachian	Arkansas North Louisiana Mississippi Montana	Onshore Louisiana Offshore Louisiana Total Louisiana Southeast New Mexico Total New Mexico North Dakota Oklahoma	East Texas Gulf Coast Texas Panhandle Texas Southwest Texas Utah Wyoming
(1, 560)	(1, 893)	(1, 1937)	0	0	(1, 492)	(4, 1466)	(10, 5953)	

1.0	1.0+ to 1.2		1.1 to 1.2	1.7 to 1.8	3.3 to 3.4	8.8 to 8.9
South Louisiana	Alabama Alaska Onshore California Total California	Northwest New Mexico North Central Texas West Texas Total Texas	Nebraska	Illinois	Indiana	Michigan
(1, 706)	(6, 4425)		(1, 616)	(1, 1062)	(1, 566)	(1, 308)

Total less than 1	Total 1	Total greater than 1	GRAND TOTAL
(18, 11301)	(1, 706)	(10, 6977)	(29, 18984)

TABLE 5D – JAS DATA AS PUBLISHED versus JAS SAMPLE – 1959
Distribution of Y_1/Y_0 at 5,000 feet: PRODUCTIVE WELLS

.8 to .9	.9 to 1.0−	1.0	1.0+ to 1.1	
Arkansas Michigan	Appalachian Total California Illinois Kentucky Total Louisiana Gulf Coast Texas Southwest Texas West Texas	Onshore California Offshore California	Alaska Colorado Indiana Kansas South Louisiana Onshore Louisiana Offshore Louisiana Montana Northwest New Mexico	Southeast New Mexico Total New Mexico Oklahoma North Central Texas Panhandle Texas Total Texas Utah Wyoming
(2, 798)	(6, 12046)	(2, 958)	(14, 13970)	

1.1 to 1.2	1.2 to 1.3	1.3 to 1.4	1.4 to 1.5	1.5 to 1.6
North Dakota East Texas		Mississippi		North Louisiana
(2, 1091)	0	(1, 283)	0	(1, 916)

Total less than 1	Total 1	Total greater than 1	GRAND TOTAL
(8, 12844)	(2, 958)	(18, 16260)	(28, 30062)

TABLE 5E – JAS DATA AS PUBLISHED versus JAS SAMPLE – 1959
Distribution of Y_1/Y_0 at 10,000 feet: PRODUCTIVE WELLS

.3 to .4	.4 to .5	.5 to .6	.6 to .7	.7 to .8	.8 to .9	.9 to 1.0−
Kansas			Michigan	Kentucky North Central Texas	North Louisiana	Arkansas Total California Colorado Illinois Total Louisiana Northwest New Mexico Gulf Coast Texas Total Texas Utah
(1, 1943)	0	0	(1, 295)	(2, 5733)	(1, 916)	(6, 3848)

1.0	1.0+ to 1.1			1.1 to 1.2
Oklahoma	Alaska Onshore California Offshore California Indiana South Louisiana Onshore Louisiana	Offshore Louisiana Montana Southeast New Mexico Total New Mexico North Dakota	East Texas Panhandle Texas Southwest Texas West Texas Wyoming	Appalachian Mississippi
(1, 3175)		(14, 11806)		(2, 2346)

Total less than 1	Total 1	Total greater than 1	GRAND TOTAL
(11, 12735)	(1, 3175)	(16, 14152)	(28, 30062)

TABLE 5F – JAS DATA AS PUBLISHED versus JAS SAMPLE – 1959
Distribution of Y_1/Y_0 at 15,000 feet: PRODUCTIVE WELLS

0 to .1	.1 to .2	.2 to .3	.3 to .4	.4 to .5	.5 to .6	.6 to .7	.7 to .8	.8 to .9
Kansas				North Louisiana North Central Texas	Michigan	Kentucky	Colorado	
(1, 1943)	0	0	0	(2, 4214)	(1, 295)	(1, 2435)	(1, 248)	0

.9 to 1.0−		1.0	1.0+ to 1.1		1.1 to 1.2	1.2 to 1.3
Illinois South Louisiana Onshore Louisiana Total Louisiana Mississippi Montana Northwest New Mexico	Southeast New Mexico Total New Mexico North Dakota Oklahoma East Texas Total Texas Utah		Alaska Arkansas Onshore California Offshore California Total California Indiana	Offshore Louisiana Gulf Coast Texas Southwest Texas West Texas Wyoming	Panhandle Texas	Appalachian
(10, 8748)		0	(10, 9077)		(1, 1039)	(1, 2063)

Total less than 1	Total 1	Total greater than 1	GRAND TOTAL
(16, 17883)	0	(12, 12179)	(28, 30062)

89

TABLE 5G – JAS DATA AS PUBLISHED versus JAS SAMPLE – 1959
Distribution of Y_1/Y_0 at 5,000 feet: OIL WELLS

.5 to .6	.6 to .7	.7 to .8	.8 to .9	.9 to 1.0-	1.0
Appalachian			Total Louisiana	Arkansas Illinois Michigan Montana Southwest Texas West Texas Total Texas	Offshore California Offshore Louisiana
(1, 881)	0	0	0	(6, 6832)	(2, 276)

1.0+ to 1.1			1.1 to 1.2	1.2 to 1.3	1.3 to 1.4	1.4 to 1.5
Onshore California Total California Colorado Indiana Kansas Kentucky North Louisiana	South Louisiana Onshore Louisiana Northwest New Mexico	Southeast New Mexico Total New Mexico Oklahoma North Central Texas Utah Wyoming	North Dakota Gulf Coast Texas Panhandle Texas		East Texas	Mississippi
		(13, 14351)	(3, 1716)	0	(1, 715)	(1, 251)

Total less than 1	Total 1	Total greater than 1	GRAND TOTAL
(7, 7713)	(2, 276)	(18, 17033)	(27, 25022)

TABLE 5H – JAS DATA AS PUBLISHED versus JAS SAMPLE – 1959
Distribution of Y_1/Y_0 at 10,000 feet: OIL WELLS

.1 to .2	.2 to .3	.3 to .4	.4 to .5	.5 to .6	.6 to .7	.7 to .8	.8 to .9
Appalachian					Michigan		Arkansas Kansas Total Louisiana
(1, 881)	0	0	0	0	(1, 241)	0	(2, 2222)

.9 to 1.0-	1.0	1.0+ to 1.1			1.1 to 1.2
Colorado Illinois North Louisiana Oklahoma East Texas Gulf Coast Texas Panhandle Texas	Total California	Onshore California Offshore California Indiana Kentucky South Louisiana Onshore Louisiana Offshore Louisiana	Montana Northwest New Mexico Southeast New Mexico	Total New Mexico North Dakota North Central Texas West Texas Total Texas Utah Wyoming	Mississippi Southwest Texas
(7, 6756)	0			(14, 13754)	(2, 1168)

Total less than 1	Total 1	Total greater than 1	GRAND TOTAL
(11, 10100)	0	(16, 14922)	(27, 25022)

TABLE 5I – JAS DATA AS PUBLISHED versus JAS SAMPLE – 1959
Distribution of Y_1/Y_0 at 15,000 feet: OIL WELLS

0 to .1	.1 to .2	.2 to .3	.3 to .4	.4 to .5	.5 to .6	.6 to .7	.7 to .8	.8 to .9
Appalachian				Michigan		East Texas	Kansas Panhandle Texas	Arkansas Colorado Mississippi North Dakota Gulf Coast Texas
(1, 881)	0	0	0	(1, 241)	0	(1, 715)	(2, 2522)	(5, 1827)

.9 to 1.0–		1.0	1.0+ to 1.1			1.1 to 1.2	1.2 to 1.3	1.3 to 1.4	1.4 to 1.5
Onshore California Total California Illinois North Louisiana	Total Louisiana Oklahoma North Central Texas Utah		Offshore California Indiana Kentucky South Louisiana	Onshore Louisiana Offshore Louisiana Montana	N.W. New Mexico S.E. New Mexico Total New Mexico West Texas Total Texas Wyoming				Southwest Texas
(6, 8571)		0	(10, 9348)			0	0	0	(1, 917)

Total less than 1	Total 1	Total greater than 1	GRAND TOTAL
(16, 14757)	0	(11, 10265)	(27, 25022)

TABLE 5J – JAS DATA AS PUBLISHED versus JAS SAMPLE – 1959
Distribution of Y_1/Y_0 at 5,000 feet: GAS WELLS

.8 to .9	.9 to 1.0–	1.0	1.0+ to 1.1		1.1 to 1.2
Mississippi	Onshore California Total California Kentucky Northwest New Mexico Oklahoma Gulf Coast Texas Panhandle Texas Southwest Texas		Appalachian Arkansas Colorado Kansas South Louisiana Onshore Louisiana	Southeast New Mexico Total New Mexico East Texas North Central Texas West Texas Wyoming	Offshore Louisiana
(1, 32)	(7, 2284)	0	(10, 2371)		(1, 74)

1.2 to 1.3	1.3 to 1.4	1.4 to 1.5	1.5 to 1.6	1.6 to 1.7	1.7 to 1.8
					North Louisiana
0	0	0	0	0	(1, 164)

Total less than 1	Total 1	Total greater than 1	GRAND TOTAL
(8, 2316)	0	(12, 2609)	(20, 4925)

91

TABLE 5K – JAS DATA AS PUBLISHED versus JAS SAMPLE – 1959
Distribution of Y_1/Y_0 at 10,000 feet: GAS WELLS

.8 to .9	.9 to 1.0 –		1.0
Kansas Kentucky Southeast New Mexico Total New Mexico	Appalachian Onshore California Total California North Louisiana Mississippi Northwest New Mexico Oklahoma Gulf Coast Texas North Central Texas Southwest Texas Total Texas		South Louisiana Onshore Louisiana
(3, 522)	(9, 3301)		(1, 319)

1.0 + to 1.1		1.1 to 1.2
Arkansas Colorado Offshore Louisiana Total Louisiana	East Texas West Texas Wyoming	Panhandle Texas
(6, 506)		(1, 277)

Total less than 1	Total 1	Total greater than 1	GRAND TOTAL
(12, 3823)	(1, 319)	(7, 783)	(20, 4925)

TABLE 5L – JAS DATA AS PUBLISHED versus JAS SAMPLE – 1959
Distribution of Y_1/Y_0 at 15,000 feet: GAS WELLS

.4 to .5	.5 to .6	.6 to .7	.7 to .8	.8 to .9	.9 to 1.0 –	
North Louisiana		Kansas Total New Mexico	Kentucky Southeast New Mexico	North Central Texas Total Texas	Appalachian Arkansas Onshore California Total California Colorado South Louisiana	Onshore Louisiana Total Louisiana Northwest New Mexico Oklahoma East Texas West Texas
(1, 164)	0	(1, 183)	(2, 339)	(1, 205)	(9, 2805)	

1.0	1.0 + to 1.1	1.1 to 1.2	1.2 to 1.3	1.3 to 1.4
Gulf Coast Texas	Offshore Louisiana Mississippi Southwest Texas Wyoming			Panhandle Texas
(1, 412)	(4, 540)	0	0	(1, 277)

Total less than 1	Total 1	Total greater than 1	GRAND TOTAL
(14, 3696)	(1, 412)	(5, 817)	(20, 4925)

92

7. Results: Changes in the Parameters, 1955-56 to 1959

The detailed comparisons discussed in this section are shown in Appendix B, Table B5. Throughout, the subscript 1 denotes the later year and the subscript 0 the earlier year of the comparison. For the reasons given in the preceding section, little reliance can be placed on the comparisons for individual data areas. Probably they are most reliable for those areas in which the *JAS* published figures were shown to be least biased. Be that as it may, it is the over-all picture which merits our attention in this final section.

Two details of procedure should be explained before we look at our results. First, in order to separate changes in real cost from changes in money cost induced by price changes only, K and all cost figures[41] for 1955 and 1956 were adjusted to 1959 prices before making the comparisons. As in the first of the two studies appearing in this volume, the index used was the Independent Petroleum Association of America's index of drilling costs. The revised index of items purchased by operators unadjusted for depth was combined with the unrevised index of contractor rates, using the 1959 weights.[42]

Second, the tables presented in this section should be read as rough graphs in a manner that is illustrated in Figure 5. In each table, the rows correspond to changes from 1955 to 1959 and the columns to changes from 1956 to 1959. Accordingly, each table is partitioned by vertical and horizontal lines through the point of no change.[43] If the 1955–1959 and 1956–1959 comparisons both show an increase, the resulting entry in the table will appear to the northeast of the intersection of these lines; if they both show a decrease, it will appear to the southwest; if 1955–1959 shows an increase and 1956–1959 a decrease, the entry will appear to the northwest; and if the reverse is true, it will appear to the southeast.[44] Moreover, if both comparisons show roughly the same order of magnitude of increase or decrease, the entry will appear close to a diagonal running from lower left to upper right. On the other hand, if one comparison shows a decrease of roughly the same magnitude as that of an increase shown by the other, the entry will lie close to a diagonal running from upper left to lower right.[45] It is reassuring to

---

[41] α is independent of price levels.

[42] See p. 10 for further detail. As to the effect of adjustment by a national index on our comparisons of less aggregative figures, there seems no reason to think that any problems here would be more than minor.

[43] Zero for $\alpha_1 - \alpha_0$ and 1.0 for all the other comparisons which are in terms of ratios.

[44] Exceptions to this are the separate comparisons for Onshore and Offshore California, for which 1955 data are not separately available, and the case of K_1/K_0 for Nebraska, where the 1955–1959 comparison is indeterminate, since both estimates are linear as described in Appendix A.

[45] In the tables themselves, these diagonals are indicated by the heavily outlined boxes.

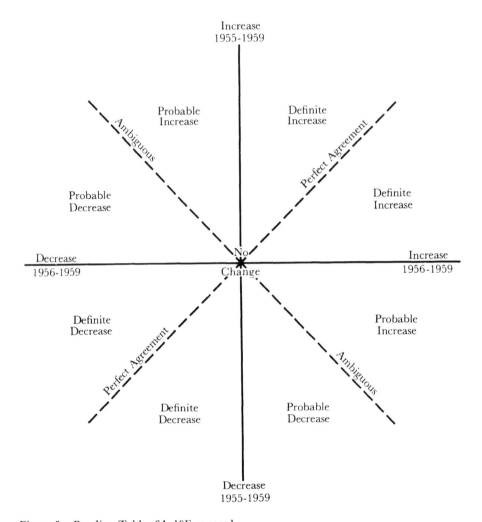

Figure 5.   Reading Tables 6A–10F as graphs.

note that the latter case happens rather infrequently and, indeed, that both comparisons generally show the same direction of change.[46]

Thus the tables can be used to gain a visual impression of the changes found. That impression may be summarized in the following rough way.

[46] This is especially reassuring in view of the fact that Table B5 of Appendix B shows a number of cases in which the results of the 1956 survey appear rather different from those of the 1955 one. In such cases, 1955 tends to appear closer to 1959 than does 1956, both in terms of parameter estimates and in terms of unexplained variance and asymptotic standard errors (see Table B3). There is no apparent reason why this should be the case; fortunately the comparisons with 1959 (as stated) come out roughly the same whether 1955 or 1956 is used as an initial year, so that this phenomenon is not particularly important for our purposes. Its existence does point up the danger of drawing conclusions from the *JAS* as published for any one year and specific area.

An entry for which both 1955–1959 and 1956–1959 show an increase is counted as a definite increase over the full period, and similarly for decreases. An entry for which the two comparisons show opposite directions of change but for which the indicated increase exceeds the indicated decrease by more than a certain absolute amount (.25 in the case of K_1/K_0, .25 × 10⁻⁴ in the case of $\alpha_1 - \alpha_0$, and .1 in the case of all other comparisons) is counted as a probable increase; if the indicated decrease exceeds the indicated increase by more than that amount, the entry is counted as a probable decrease. Finally, if the two comparisons show opposite directions of change and do not differ in magnitude by more than the stated amount, the entry is counted as ambiguous. Thus, as is shown in Figure 5, probable increases lie above the upper left—lower right diagonal and probable decreases below it, while ambiguous cases lie approximately on that diagonal. As in Section 6, the totals for the categories are given in terms of number of data areas and number of wells drilled therein in 1959,[47] excluding those data areas which are aggregates of other included ones.

The results are presented as follows. Tables 6A and 6B (pages 98–99) give the comparisons for K_1/K_0 for dry and productive wells, respectively; Tables 7A and 7B (pages 100–101) give them for $\alpha_1 - \alpha_0$. The results for H_1/H_0 are presented in Tables 8A and 8B (pages 102–3). Finally, comparisons of marginal cost at 5,000, 10,000, and 15,000 feet are given in Tables 9A–9F (pages 104–9) and comparisons of total costs at the same depths are given in Tables 10A–10F (pages 110–15). As in Section 6, note that the results on H_1/H_0 are the appropriate ones for comparisons of both marginal and total costs for very shallow wells.

A summary by number of data areas and wells drilled is given in Tables 11A and 11B (page 116).

Beginning our discussion with changes in α and K, there seems to be no clear tendency for either parameter to rise rather than fall, the results being fairly evenly distributed over increases and decreases for both parameters and for both dry and productive wells. There is some indication that increases in α predominated in areas with large amounts of drilling. The implications appear to be that the net effects of technological and other changes over the period in question were depth neutral or somewhat shallowness-favoring if we measure them in this rough way. This does not mean that such was the case for any particular area, merely that areas with depth-favoring changes appear to have been about as numerous as areas with shallowness-favoring changes, although less important in terms of wells drilled. As we shall see when we come to discuss the behavior of marginal and total costs, there is considerable evidence that the *importance* of depth-favoring changes was less during this period than that of shallowness-favoring ones. Moreover, the fact that the *JAS* for 1959 is known to under-

---

[47] The distribution of drilling changed over the period in question but not so much as substantially to alter the rough summary being described.

state α, especially when it is large, means that there is a tendency for observed changes in α to be biased downward, since the 1955–56 survey did not use freehand curve fitting. It follows that even an equal observed frequency of increases and decreases in α may correspond to a true preponderance of increases, and hence our evidence points toward shallowness-favoring changes.

We turn now to the behavior of marginal and total drilling costs themselves. On the whole, the evidence is rather stronger for cost reduction in the case of dry than in the case of productive wells. As the results for the latter are somewhat suspect, since a change in the mix of oil and gas wells in the *JAS* sample can easily distort them, we should take the dry well evidence as more important; however, the general indications are the same in both cases.

Recalling that changes in H can be treated as changes in both marginal and total cost at very shallow depths (see pages 75–76), we may use the observed changes in that parameter to round out the picture afforded by our marginal and total cost results at 5,000, 10,000, and 15,000 feet.

For dry wells, the results are quite clear. There is strong evidence of reduction in marginal cost below 10,000 feet. At about 10,000 feet there are still more areas experiencing a decrease in marginal cost than experiencing an increase, but the distribution of drilling is about even. By 15,000 feet, things have clearly swung the other way in terms of amount of drilling, and increases slightly predominate over decreases in terms of number of data areas.

For total costs in dry wells, the picture is similar. Even at 10,000 feet the evidence is overwhelmingly in favor of total cost decreases, but by 15,000 feet things are about even in terms of amount of drilling and not nearly so one-sided as at lesser depths in terms of number of areas. As in the case of marginal costs, therefore, decreases seem to have occurred in most areas in all but the very deep wells; at great depth, such decreases are considerably less pronounced. The results on both marginal and total costs, then, imply that the magnitude of depth-favoring changes was rather less than that of shallowness-favoring ones. Hence, it would seem that the tendency discussed in Section 3 for technological change to move to more costly structures has not been so clear in the case of very deep wells as it has in the case of relatively shallow ones. In spite of this, the over-all impression is clearly one of general cost reduction over the period studied, and the reductions are occasionally of impressive magnitude, as can be seen from the tables in this section and from Table B5 of Appendix B.

Turning to the case of productive wells, where the evidence for cost reduction is less clear-cut than in the case of dry wells (but should be given rather less weight), a similar pattern emerges. At less than about 5,000 feet, increases and decreases in marginal cost seem about equally matched; at greater depths increases predominate, but the predominance is not extreme. Since a shallowness-favoring change which reduces total costs can readily produce increases in marginal cost at all but the smallest depths, the evidence is once again in favor of shallowness-favoring change.

This is borne out by the results for total costs in productive wells. As in the case of dry wells, at less than 10,000 feet, the evidence is largely in favor of cost decreases. At 10,000 feet, where we found overwhelming evidence of cost decreases in the dry well case, the results are rather even for productive wells: decreases are slightly predominant in terms of number of data areas, and increases are slightly ahead in terms of the number of wells drilled. By 15,000 feet, the results have clearly, but not overwhelmingly, swung to the side of cost increases. In general, then, the picture is one of shallowness-favoring change and of cost reduction at depths to which the overwhelming majority of wells are drilled. This reinforces the much stronger evidence in the case of dry wells.

To summarize, we have found considerable evidence of cost reductions in the 1955–56–1959 period. That evidence is stronger for dry than for productive wells and points toward greater reductions in relatively shallow than in very deep wells. The fact that the *JAS* data as published are subject to large biases makes it pointless to discuss the precise magnitudes of the cost changes indicated; nevertheless, the over-all picture seems clear. Technological change in the period 1955–1959 obviously more than kept pace in most areas with the tendency discussed in Section 3 to move toward more costly structures. Indeed, it is more than likely that average real drilling costs per well fell in most areas, as the increases in depth during the period were not enough to compensate for the technologically induced downward shifts in the cost function.

Thus, if the domestic petroleum industry experienced increasing costs in this period—and it is by no means clear that this was the case—the increases came either from price changes (an effect not special to extractive industries) or from changes in the costs of petroleum discovery, development, and production (other than well drilling costs), all of which are outside the province of this study.[48]

So far as real drilling costs themselves are concerned, the general experience of the industry during the period was one of declining, not increasing, costs. The effects of future technological change on future costs are, of course, uncertain, so that it is not obvious that the trend toward lower costs will continue. One thing is clear, however: costs do not necessarily increase over time as wells get deeper, even though the data for any one year may show a sharp increase of drilling costs with depth.[49]

[48] Such effects include possible shifts in the geographic distribution of drilling toward high-cost data areas induced by fuller exploitation of low-cost ones. This is a complicated matter, but there is certainly little evidence of a large shift of this type *within* data areas.

[49] To draw the opposite conclusion on such a basis is especially dangerous if the data examined apply to the United States as a whole or to any large aggregate area.

Change 1955-59 \ Change 1956-59	0 to .25	.25 to .5	.5 to .75	.75 to 1.0	1.0	1.0+ to 1.25	1.25 to 1.5	1.5 to 1.75	1.75 to 2.0	Greater than 2.0
Greater than 2.0	Appalachian						Mississippi			Alabama, Colorado,c Indiana, Nebraska Panhandle Texas
1.75 to 2.0										Arkansas
1.5 to 1.75		Michigan								North Dakota
1.25 to 1.5							Total New Mexico			
1.0+ to 1.25			East Texas							
1.0										
.75 to 1.0–				North Central Texas						
.5 to .75		Montana, Gulf Coast, Texas, Utah	Total California, Oklahoma, Total Texas, Wyoming	Southwest Texas		Offshore Louisiana	Kansas			Kentucky
.25 to .5		West Texas	Total Louisiana							
.0 to .25	Illinois a,b		Onshore Louisiana	Onshore California						
Not available										

NOTE: Tables 6A – 10F may be read as graphs. See Figure 5 and pp. 93-96.
a Estimate for 1955 linear and unreliable (see Appendix A).
b Estimate for 1956 linear and unreliable (see Appendix A).
c Estimate for 1959 linear and unreliable (see Appendix A).

TABLE 6B – CHANGES IN THE DRILLING COST FUNCTION: 1955-56 – 1959

Distribution of K_1/K_0: PRODUCTIVE WELLS

Change 1956-59 \ Change 1955-59	0 to .25	.25 to .5	.5 to .75	.75 to 1.0 –	1.0	1.0+ to 1.25	1.25 to 1.5	1.5 to 1.75	1.75 to 2.0	Greater than 2.0
Indeterminate										Nebraska[a,c]
Greater than 2.0						Panhandle Texas		Southwest Texas		Indiana[c] Montana
1.75 to 2.0	Total California	Michigan		Colorado				North Central Texas		
1.5 to 1.75			Arkansas	North Dakota			Appalachian			
1.25 to 1.5										
1.0+ to 1.25						Total New Mexico				
1.0				Wyoming		Oklahoma	Kansas, Onshore Louisiana	Total Louisiana		Offshore Louisiana
.75 to 1.0 –		Mississippi	East Texas, Gulf Coast Texas, West Texas, Total Texas							
.5 to .75	Utah[b]									
.25 to .5										Illinois[a]
0 to .25	Kentucky[a,b]									
Not available	Onshore California, Offshore California									

[a]Estimate for 1955 linear and unreliable (see Appendix A).
[b]Estimate for 1956 linear and unreliable (see Appendix A).
[c]Estimate for 1959 linear and unreliable (see Appendix A).

99

TABLE 7A – CHANGES IN THE DRILLING COST FUNCTION: 1955-56 – 1959; Distribution of $\alpha_1 - \alpha_0$: DRY WELLS

($\times 10^{-4}$ omitted in all figures)

Change 1955-59 ↓ \ Change 1956-59 →	Greater than 1.5	1.25 to 1.5	1.0 to 1.25	.75 to 1.0	.5 to .75	.25 to .5	0+ to .25	0	−.25 to 0−	−.5 to −.25	−.75 to −.5	−1.0 to −.75	−1.25 to −1.0	−1.5 to −1.25	Less than −1.5
Greater than 1.5	Illinois[a,b]														
1.25 to 1.5															
1.0 to 1.25			West Texas												
.75 to 1.0				Gulf Coast Texas											
.5 to .75							Onshore La.				Kans.				
.25 to .5				Mont.	Okla.	Wyo.									
0+ to .25					Utah		Total La.								
0							S.W. Tex., Total Texas								
−.25 to 0−					Mich.	N. Cen. Texas	Total Calif.		Total N. Mex.						
−.5 to −.25					East Texas					Offshore La.					
−.75 to −.5										Miss.					
−1.0 to −.75	Appala-chian									Ark.					
−1.25 to −1.0													Pan. Texas		
−1.5 to −1.25													Ala.	Colo.	
Less than −1.5														Ind.	Ky., Nebr.[c] N. Dak.
Not available									Onshore Calif.						

[a] Estimate for 1955 linear and unreliable (see Appendix A).
[b] Estimate for 1956 linear and unreliable (see Appendix A).
[c] Estimate for 1959 linear and unreliable (see Appendix A).

TABLE 7B – CHANGES IN THE DRILLING COST FUNCTION: 1955-56 – 1959 Distribution of $\alpha_1 - \alpha_0$: PRODUCTIVE WELLS ($\times 10^{-4}$ omitted in all figures)

Change 1955-59 \ Change 1956-59	Greater than 1.5	1.25 to 1.5	1.0 to 1.25	.75 to 1.0	.5 to .75	.25 to .5	0+ to .25	0	-.25 to 0-	-.5 to -.25	-.75 to -.5	-1.0 to -.75	-1.25 to -1.0	-1.5 to -1.25	Less than -1.5	Not available
Greater than 1.5																
1.25 to 1.5													Illinois[a]			
1.0 to 1.25																
.75 to 1.0																
.5 to .75					East Central Texas	Tot. Tex., Ky.[a,b], E.Tex., W. Tex.	Okla									
.25 to .5	Utah[b]			Miss.	Total Calif.				Wyo.	Onshore La., Total La	Kans.					
0+ to .25											N. Cen. Texas					
0															Nebr.[a,c]	Ind.[c]
-.25 to 0-						N. Dak.			Appala., Tot. N. Mex.							
-.5 to -.25					Ark., Colo.						S. W. Texas	Offshore La				
-.75 to -.5									Pan. Texas							
-1.0 to -.75																
-1.25 to -1.0			Mich.													
-1.5 to -1.25														Mont.		
Not available	Offshore Calif.				Onshore Calif.											

[a] estimate for 1955 linear and unreliable (see Appendix A).
[b] estimate for 1956 linear and unreliable (see Appendix A).
[c] estimate for 1959 linear and unreliable (see Appendix A).

101

TABLE 8A – CHANGES IN THE DRILLING COST FUNCTION: 1955-56 – 1959
Distribution of H_1/H_0: DRY WELLS

Change 1955-59 \ Change 1956-59	.2 to .3	.3 to .4	.4 to .5	.5 to .6	.6 to .7	.7 to .8	.8 to .9	.9 to 1.0–	1.0	1.0+ to 1.1	1.1 to 1.2	1.2 to 1.3	1.3 to 1.4	1.4 to 1.5	1.5 to 1.6	1.6 to 1.7	1.7 to 1.8	1.8 to 1.9	1.9 to 2.0	Greater than 2.0
Greater than 2.0													Appalachian							Colo.
1.9 to 2.0																				
1.8 to 1.9													Ark.							
1.7 to 1.8													Pan. Tex.							
1.6 to 1.7																				
1.5 to 1.6																				
1.4 to 1.5															Ala.					
1.3 to 1.4																				
1.2 to 1.3										Miss.	Total N. Mex.									N. Dak.
1.1 to 1.2										Kans.	Ky.									
1.0+ to 1.1																				
1.0																				
.9 to 1.0–						East Tex.	N. Cen. Tex.	Mich.												
.8 to .9						Okla.	Total Tex.													Nebr.[c]
.7 to .8				Utah		Total Calif., Wyo.	S.W. Tex.													
.6 to .7				E. Cen. Tex.	Ill.,[a,b] Mont.			Offshore La.												
.5 to .6				Ind.		Total La.														
.4 to .5			West Tex.		Onshore La.															
.3 to .4							Onshore Calif.													
.2 to .3																				
Not available																				

[a] Estimate for 1955 linear and unreliable (see Appendix A).
[b] Estimate for 1956 linear and unreliable (see Appendix A).
[c] Estimate for 1959 linear and unreliable (see Appendix A).

TABLE 8B – CHANGES IN THE DRILLING COST FUNCTION, 1955-56 – 1959

Distribution of H_1/H_0: PRODUCTIVE WELLS

Change 1955-59 ＼ Change 1956-59	.3 to .4	.4 to .5	.5 to .6	.6 to .7	.7 to .8	.8 to .9	.9 to 1.0–	1.0	1.0+ to 1.1	1.1 to 1.2	1.2 to 1.3	1.3 to 1.4	1.4 to 1.5	1.5 to 1.6	1.6 to 1.7	1.7 to 1.8	1.8 to 1.9	1.9 to 2.0	Greater than 2.0
Greater than 2.0															Ind.[c]				
1.9 to 2.0										Offshore La.									
1.8 to 1.9																			
1.7 to 1.8									Colo.										
1.6 to 1.7									Mich.										
1.5 to 1.6																			
1.4 to 1.5							N. Dak., Pan. Tex.						Mont., S.W. Tex.						
1.3 to 1.4																			
1.2 to 1.3							Ark.			Appala., N-C Tex.									
1.1 to 1.2																			
1.0+ to 1.1					Total Calif.														Nebr.[c]
1.0																			
.9 to 1.0–						West Tex.	Total N. Mex.		Kans.		Total La.								
.8 to .9		Miss.				Total Tex.	Wyo.		Okla.	Onshore La.									
.7 to .8				Glf. Cst. Tex.															
.6 to .7	Utah[b]				East Tex.				Ky.[a,b]			Illinois[a]							
.5 to .6																			
.4 to .5																			
.3 to .4																			
Not available		Offshore Calif.			Onshore Calif.														

[a] Estimate for 1955 linear and unreliable (see Appendix A).
[b] Estimate for 1956 linear and unreliable (see Appendix A).
[c] Estimate for 1959 linear and unreliable (see Appendix A).

TABLE 9A – CHANGES IN THE DRILLING COST FUNCTION: 1955-56 – 1959
Distribution of MC_1/MC_0 at 5,000 Feet: DRY WELLS

Change 1955-59 \ Change 1956-59	.2 to .3	.3 to .4	.4 to .5	.5 to .6	.6 to .7	.7 to .8	.8 to .9	.9 to 1.0−	1.0	1.0+ to 1.1	1.1 to 1.2	1.2 to 1.3	1.3 to 1.4	1.4 to 1.5	1.5 to 1.6	1.6 to 1.7	1.7 to 1.8	1.8 to 1.9	1.9 to 2.0	Greater than 2.0
Greater than 2.0																				Appalachian
1.9 to 2.0																				
1.8 to 1.9																			Ill. a,b	
1.7 to 1.8																				
1.6 to 1.7																				
1.5 to 1.6											Colo.									
1.4 to 1.5																				
1.3 to 1.4																				
1.2 to 1.3																				
1.1 to 1.2												Mich.								
1.0+ to 1.1						Kans.	Wyo.			Ark., Total, New Mex.										
1.0																				
.9 to 1.0−							Tot. Tex.; G.C.Tex., S.W.Tex., W.Tex.; Miss., Mont., E.Tex., N-C Tex.; Tot. Calif.	Okla.												
.8 to .9	Ind.		Ky.		Utah						N. Dak.									
.7 to .8				Nebr. c		Pan. Tex.; Offshr. La., Tot.La.														
.6 to .7						Onshr. La.														
.5 to .6								Ala.												
.4 to .5																				
.3 to .4																				
.2 to .3																				
Not available							Onshr. Calif.													

a Estimate for 1955 linear and unreliable (see Appendix A).
b Estimate for 1956 linear and unreliable (see Appendix A).
c Estimate for 1959 linear and unreliable (see Appendix A).

TABLE 7B – CHANGES IN THE DRILLING COST FUNCTION, 1955-56 – 1959

Distribution of MC_1/MC_0 at 10,000 Feet: DRY WELLS

Change 1955-59 \ Change 1956-59	>2.0	1.9–2.0	1.8–1.9	1.7–1.8	1.6–1.7	1.5–1.6	1.4–1.5	1.3–1.4	1.2–1.3	1.1–1.2	1.0+–1.1	1.0	.9–1.0−	.8–.9	.7–.8	.6–.7	.5–.6	.4–.5	.3–.4	.2–.3	.1–.2	0–.1
Greater than 2.0	Ill.[a,b]																					
1.9 to 2.0																						
1.8 to 1.9																						
1.7 to 1.8																						
1.6 to 1.7																						
1.5 to 1.6																						
1.4 to 1.5								West Tex.						Onshore La.			Kans.				Ind.	
1.3 to 1.4													Wyo.									
1.2 to 1.3																						
1.1 to 1.2							Mont.		Glf.Cst. Tex.													Nebr.[c]
1.0+ to 1.1	Appalachian									Okla.			Total Tex.									
1.0																						
.9 to 1.0−				Mich.									S.W. Tex.								Ky.	
.8 to .9													Tot. New Mex., Utah		Total La.							
.7 to .8										East Tex., N-C Tex.				Total Calif.		Ofshr. La., N. Dak.						
.6 to .7																						
.5 to .6																						
.4 to .5														Miss.	Ark.							
.3 to .4																						
.2 to .3																	Ala., Colo.	Pan. Tex.				
.1 to .2																						
0 to .1																						
Not available														Onshore Calif.								

a Estimate for 1955 linear and unreliable (see Appendix A).
b Estimate for 1956 linear and unreliable (see Appendix A).
c Estimate for 1959 linear and unreliable (see Appendix A).

TABLE 9C – CHANGES IN THE DRILLING COST FUNCTION: 1955-56 – 1959
Distribution of MC_1/MC_0 at 15,000 Feet: DRY WELLS

Change 1955-59 \ Change 1956-59	0 to .1	.1 to .2	.2 to .3	.3 to .4	.4 to .5	.5 to .6	.6 to .7	.7 to .8	.8 to .9	.9 to 1.0-	1.0	1.0+ to 1.1	1.1 to 1.2	1.2 to 1.3	1.3 to 1.4	1.4 to 1.5	1.5 to 1.6	1.6 to 1.7	1.7 to 1.8	1.8 to 1.9	1.9 to 2.0	Greater than 2.0
Greater than 2.0	Ind.									Onshr. La.												Ill.,a,b W. Tex.
1.9 to 2.0																						
1.8 to 1.9					Kans.								Wyo.									
1.7 to 1.8																						
1.6 to 1.7																						
1.5 to 1.6																				Glf. Cst. Tex.		
1.4 to 1.5																						Mont.
1.3 to 1.4																	Okla.					
1.2 to 1.3								Total La.														
1.1 to 1.2										S. W. Tex.		Total Tex.										
1.0+ to 1.1																						
1.0													Utah									
.9 to 1.0-																						Mich.
.8 to .9	Ky.					Offshr. La.																Appala-chian
.7 to .8	Nebr.c									Total Calif.				N. Cen. Tex.								
.6 to .7										Total N. Mex.							East Tex.					
.5 to .6							Ark.															
.4 to .5								Miss.														
.3 to .4			N. Dak.																			
.2 to .3																						
.1 to .2			Pan. Tex.	Ala.																		
0 to .1				Colo.																		
Not available																						

a Estimate for 1955 linear and unreliable (see Appendix A).
b Estimate for 1956 linear and unreliable (see Appendix A).
c Estimate for 1959 linear and unreliable (see Appendix A).

TABLE 30 CHANGES IN THE DRILLING COST FUNCTION, 1955-56 – 1959

Distribution of MC_1/MC_0 at 5,000 Feet: PRODUCTIVE WELLS

Change 1955-59 \ Change 1956-59	.2 to .3	.3 to .4	.4 to .5	.5 to .6	.6 to .7	.7 to .8	.8 to .9	.9 to 1.0-	1.0	1.0+ to 1.1	1.1 to 1.2	1.2 to 1.3	1.3 to 1.4	1.4 to 1.5	1.5 to 1.6	1.6 to 1.7	1.7 to 1.8	1.8 to 1.9	1.9 to 2.0	Greater than 2.0
Greater than 2.0																				
1.9 to 2.0																				
1.8 to 1.9																				
1.7 to 1.8																				
1.6 to 1.7						Ill.[a]														
1.5 to 1.6																				
1.4 to 1.5													Colo.					Mich.		
1.3 to 1.4																				
1.2 to 1.3												Ark.								
1.1 to 1.2								Offshr. La.		West Tex.	Ky,[a,b]									
1.0+ to 1.1						Nebr.[a,c]	Kans.				Appala., Okla.									
1.0										Tot. Calif.	N. Dak., S.W. Tex.									
.9 to 1.0-								Onshr.La., Tot. N. Mex., Wyo.		Tot. La., Tot. Tex.										
.8 to .9						Miss.	N. Cen. Tex.	Gulf Coast Tex.												
.7 to .8						Mont.	Panhandle Tex.	E. Tex.												
.6 to .7							Utah[b]													
.5 to .6																				
.4 to .5																				
.3 to .4																				
.2 to .3		Ind.[c]																		
Not available										Onshr. Calif.						Offshr. Calif.				

[a] Estimate for 1955 linear and unreliable (see Appendix A).
[b] Estimate for 1956 linear and unreliable (see Appendix A).
[c] Estimate for 1959 linear and unreliable (see Appendix A).

TABLE 9E – CHANGES IN THE DRILLING COST FUNCTION: 1955-56 – 1959
Distribution of MC_1/MC_0 at 10,000 Feet: PRODUCTIVE WELLS

Change 1956-59 → / Change 1955-59 ↓	Greater than 2.0	1.9 to 2.0	1.8 to 1.9	1.7 to 1.8	1.6 to 1.7	1.5 to 1.6	1.4 to 1.5	1.3 to 1.4	1.2 to 1.3	1.1 to 1.2	1.0+ to 1.1	1.0	.9 to 1.0-	.8 to .9	.7 to .8	.6 to .7	.5 to .6	.4 to .5	.3 to .4	.2 to .3	.1 to .2	0 to .1
Greater than 2.0							Mich.							Utahᵇ								Offshr. Calif.
1.9 to 2.0																						
1.8 to 1.9																						
1.7 to 1.8						Colo.																
1.6 to 1.7																						
1.5 to 1.6									Ark.													
1.4 to 1.5									Tot. Calif.													Onshr. Calif.
1.3 to 1.4						Ky.ᵃ,ᵇ								N. Dak.								
1.2 to 1.3									Glf. Cst. Tex.			Miss.										
1.1 to 1.2						W. Tex.	Okla.		Tot. Tex.													
1.0+ to 1.1									Appala.			E. Tex.										
1.0																						
.9 to 1.0-												Tot. N. Mex.		S.W. Tex.								
.8 to .9									Onshr. La.,Wyo.			Tot. La.				Panhandle Tex.						
.7 to .8															Offshr. La.							
.6 to .7									Kans.							N. Cen. Tex.						
.5 to .6																						
.4 to .5	Ill.ᵃ																	Mont.				
.3 to .4																						
.2 to .3									Nebr.ᵃ,ᶜ													
.1 to .2																						
0 to .1																					Ind.ᶜ	
Not available																						

ᵃEstimate for 1955 linear and unreliable (see Appendix A).
ᵇEstimate for 1956 linear and unreliable (see Appendix A).
ᶜEstimate for 1956 linear and unreliable (see Appendix A).

108

TABLE 9. CHANGES IN THE DRILLING COST FUNCTION: 1955-56 – 1959

Distribution of MC_1/MC_0 at 15,000 Feet: PRODUCTIVE WELLS

Change 1955-59 \ Change 1956-59	0 to .1	.1 to .2	.2 to .3	.3 to .4	.4 to .5	.5 to .6	.6 to .7	.7 to .8	.8 to .9	.9 to 1.0-	1.0	1.0+ to 1.1	1.1 to 1.2	1.2 to 1.3	1.3 to 1.4	1.4 to 1.5	1.5 to 1.6	1.6 to 1.7	1.7 to 1.8	1.8 to 1.9	1.9 to 2.0	Greater than 2.0
Greater than 2.0	Ill.[a]																					
1.9 to 2.0																						
1.8 to 1.9																						
1.7 to 1.8														Okla.		W. Tex.						
1.6 to 1.7																	Ky.[a,b]					
1.5 to 1.6																						
1.4 to 1.5																						
1.3 to 1.4					Kans.																	
1.2 to 1.3															Tot. Tex.			Glf. Cst. Tex.				Colo., Mich.
1.1 to 1.2									Wyo.				E. Tex.									Ark.
1.0+ to 1.1	Nebr.[a,c]						Onshr. La. / Tot. La.														Tot. Calif.	
1.0										Appala.												
.9 to 1.0-									Tot. N. Mex.													Miss.
.8 to .9																						Utah[b]
.7 to .8																						
.6 to .7																						
.5 to .6							Offshr. La.										N. Dak.					
.4 to .5								S.W. Tex. / Pan. Tex.														
.3 to .4					N. Cen. Tex.																	
.2 to .3																						
.1 to .2			Mont.																			
0 to .1	Ind.[c]																					On&Offshr. Cal.
Not available																						

[a] Estimate for 1955 linear and unreliable (see Appendix A).
[b] Estimate for 1956 linear and unreliable (see Appendix A).
[c] Estimate for 1959 linear and unreliable (see Appendix A).

TABLE 10A – CHANGES IN THE DRILLING COST FUNCTION: 1955-56 – 1959

Distribution of Y_1/Y_0 at 5,000 Feet: DRY WELLS

Change 1955-59 \ Change 1956-59	Greater than 2.0	1.9 to 2.0	1.8 to 1.9	1.7 to 1.8	1.6 to 1.7	1.5 to 1.6	1.4 to 1.5	1.3 to 1.4	1.2 to 1.3	1.1 to 1.2	1.0+ to 1.1	1.0	.9 to 1.0-	.8 to .9	.7 to .8	.6 to .7	.5 to .6	.4 to .5	.3 to .4
Greater than 2.0	Appala.						Colo.											Ind.	
1.9 to 2.0																			
1.6 to 1.7																			
1.5 to 1.6						Utah													
1.4 to 1.5															W. Tex.	Ky.			
1.3 to 1.4																			
1.2 to 1.3																			
1.1 to 1.2									Panhandle Tex.	Ark.			Miss.						
1.0+ to 1.1										Mich. Ill.[a,b]	Tot. N. Mex.		N. Cen. Tex.						
1.0											Ala.			Kans.	Tot. Calif., Mont.	Offshr. La.			
.9 to 1.0-						N. Dak.								Okla., E. Tex., S.W. Tex., Tot. Tex., Wyo.	Glf. Cst. Tex.	Tot. La.	Onshr. La.		
.8 to .9																			
.7 to .8									Nebr.[c]										
.6 to .7																			
.5 to .6																			
.4 to .5																			
.3 to .4																			
Not available														Onshr. Calif.					

[a] Estimate for 1955 linear and unreliable (see Appendix A).
[b] Estimate for 1956 linear and unreliable (see Appendix A).
[c] Estimate for 1959 linear and unreliable (see Appendix A).

TABLE 10B – CHANGES IN THE DRILLING COST FUNCTION: 1955-56 – 1959

Distribution of Y_1/Y_0 at 10,000 Feet: DRY WELLS

Change 1955-59 \ Change 1956-59	.2 to .3	.3 to .4	.4 to .5	.5 to .6	.6 to .7	.7 to .8	.8 to .9	.9 to 1.0–	1.0	1.0+ to 1.1	1.1 to 1.2	1.2 to 1.3	1.3 to 1.4	1.4 to 1.5	Greater than 2.0
Greater than 2.0	III. a,b										Appalachian				
1.4 to 1.5															
1.3 to 1.4															
1.2 to 1.3					Kans.										
1.1 to 1.2							Wyo.								
1.0+ to 1.1								West Tex.							
1.0															
.9 to 1.0–	Ind.	Ky.						Glf. Cst. Tex., Tot. Tex.		Mont., Tot. N. Mex., Okla.				Mich.	
.8 to .9	Nebr.[c]					Utah	N. Dak., S. W. Tex.			N. Central Tex.					
.7 to .8						Onshr. La., Offshr. La., Tot. La.	Tot. Calif., Colo.	Ark., Miss., East Tex.							
.6 to .7					Pan. Tex.										
.5 to .6							Ala.								
.4 to .5															
.3 to .4															
.2 to .3															
Not available							Onshore Calif.								

[a] Estimate for 1955 linear and unreliable (see Appendix A).
[b] Estimate for 1956 linear and unreliable (see Appendix A).
[c] Estimate for 1959 linear and unreliable (see Appendix A).

111

TABLE 10C – CHANGES IN THE DRILLING COST FUNCTION: 1955-56 – 1959

Distribution of Y_1/Y_0 at 15,000 Feet: DRY WELLS

Change 1955-59 \ Change 1956-59	0 to .1	.1 to .2	.2 to .3	.3 to .4	.4 to .5	.5 to .6	.6 to .7	.7 to .8	.8 to .9	.9 to 1.0–	1.0	1.0+ to 1.1	1.1 to 1.2	1.2 to 1.3	1.3 to 1.4	1.4 to 1.5	1.5 to 1.6	1.9 to 2.0	Greater than 2.0
Greater than 2.0																			Ill.[a,b]
1.9 to 2.0	Nebr.[c]																		
1.5 to 1.6						Kans.											W. Tex.		
1.4 to 1.5												Wyo.							
1.3 to 1.4							Offshr. La.		Onshr. La.										
1.2 to 1.3																Glf. Cst. Tex.	Mont.		
1.1 to 1.2													Ind.	Okla.					
1.0+ to 1.1												Tot. Tex.							
1.0										S.W. Tex.									
.9 to 1.0–		Ky.						Tot. La.	Utah	Tot. N. Mex.			N. Cen. Tex.					Mich.	Appala.
.8 to .9									Tot. Calif.										
.7 to .8					N. Dak.			Miss.						E. Tex.					
.6 to .7																			
.5 to .6									Ark.										
.4 to .5					Panhandle Tex.														
.3 to .4																			
.2 to .3					Colo.	Ala.			Onshr. Calif.										
.1 to .2																			
0 to .1																			
Not available																			

a Estimate for 1955 linear and unreliable (see Appendix A).
b Estimate for 1956 linear and unreliable (see Appendix A).
c Estimate for 1959 linear and unreliable (see Appendix A).

112

TABLE 10b – CHANGES IN THE DRILLING COST FUNCTION: 1955-56 – 1959

Distribution of Y_1/Y_0 at 5,000 Feet: PRODUCTIVE WELLS

Change 1955-59 \ Change 1956-59	.5 to .6	.6 to .7	.7 to .8	.8 to .9	.9 to 1.0-	1.0	1.0+ to 1.1	1.1 to 1.2	1.2 to 1.3	1.3 to 1.4	1.4 to 1.5	1.5 to 1.6
1.5 to 1.6									Colorado			
1.4 to 1.5											Michigan	
1.3 to 1.4												
1.2 to 1.3							Arkansas, North Dakota		Southwest Texas			
1.1 to 1.2							Montana	Appalachian				
1.0+ 1.1				Total California	Illinois[a]		Kentucky[a,b]	Nebraska[a]				
1.0												
.9 to 1.0-				Gulf Coast Texas	Kan., W.Tex., Tot.N.Mex., Tot.Tex., N.Cen. Tex., Pan. Tex.		Oklahoma	Total Louisiana				
.8 to .9		Mississippi			Wyoming		Onshore Louisiana[d]					
.7 to .8				East Texas								
.6 to .7	Utah[b]	Indiana[c]									Offshore Louisiana	
.5 to .6				Onshore California	Offshore California							
Not available												

[a] Estimate for 1955 linear and unreliable (see Appendix A).
[b] Estimate for 1956 linear and unreliable (see Appendix A).
[c] Estimate for 1959 linear and unreliable (see Appendix A).
[d] 1955-1959: .900 to 3 decimal places.

TABLE 10E – CHANGES IN THE DRILLING COST FUNCTION: 1955-56 – 1959

Distribution of Y_1/Y_0 at 10,000 Feet: PRODUCTIVE WELLS

Change 1955-59 (rows) / Change 1956-59 (cols)	Greater than 2.0	1.4 to 1.5	1.3 to 1.4	1.2 to 1.3	1.1 to 1.2	1.0+ to 1.1	1.0	.9 to 1.0-	.8 to .9	.7 to .8	.6 to .7	.5 to .6	.4 to .5	.3 to .4	.2 to .3	.1 to .2
Greater than 2.0																
1.4 to 1.5	Mich.	Colo.										III.[a]				
1.3 to 1.4			Ky.[a,b]													
1.2 to 1.3																
1.1 to 1.2			Ark.		Okla.	Appala.										
1.0+ to 1.1					N. Dak.	Tot. Calif., W. Tex., Tot. Tex.		Onshr. La., Tot. La., Tot. N. Mex., Wyo.				Nebr.[a]				
1.0																
.9 to 1.0-						Glf. Cst. Tex.			Miss.							
.8 to .9						S.W. Tex.		E. Tex.	Offshr. La.							
.7 to .8						Utah[b]				Mont., N. Cen. Tex.						
.6 to .7									Panhandle Tex.							
.5 to .6																
.4 to .5																
.3 to .4																
.2 to .3																
.1 to .2															Ind.[c]	
Not available	Offshr. Calif.					Onshr. Calif.										

[a] Estimate for 1955 linear and unreliable (see Appendix A).
[b] Estimate for 1956 linear and unreliable (see Appendix A).
[c] Estimate for 1959 linear and unreliable (see Appendix A).

TABLE 10F – CHANGES IN THE DRILLING COST FUNCTION: 1955-56 – 1959

Distribution of Y_1/Y_0 at 15,000 Feet: PRODUCTIVE WELLS

Change 1955-59 \ Change 1956-59	0 to .1	.1 to .2	.2 to .3	.3 to .4	.4 to .5	.5 to .6	.6 to .7	.7 to .8	.8 to .9	.9 to 1.0–	1.0	1.0+ to 1.1	1.1 to 1.2	1.2 to 1.3	1.3 to 1.4	1.4 to 1.5	1.5 to 1.6	1.6 to 1.7	1.7 to 1.8	1.8 to 1.9	1.9 to 2.0	Greater than 2.0
Greater than 2.0																						
1.5 to 1.6																						III.[a]
1.4 to 1.5													Okla.									
1.3 to 1.4																				Colo.		Mich.
1.2 to 1.3													W. Tex.									
1.1 to 1.2												Appala.		Ky.[a,b]								
1.0+ to 1.1		Nebr.[c]											Tot. Tex.		Glf. Cst. Tex.; Tot. Calif.		Ark.					
1.0																						
.9 to 1.0–												E. Tex.			Miss.							
.8 to .9									Tot. La.	Tot. N. Mex.		Onshr. La., Wyo.		N. Dak.								
.7 to .8																						Utah[b]
.6 to .7								Offshr. La.														
.5 to .6							N. Cen. Tex.		Panhandle Tex.	S.W. Tex.		Kans.										
.4 to .5					Mont.																	
.3 to .4																						
.2 to .3																						
.1 to .2																						
0 to .1	Ind.[c]																					
Not available															Onshr. Calif.							Offshr. Calif.

[a] Estimate for 1955 linear and unreliable (see Appendix A).
[b] Estimate for 1956 linear and unreliable (see Appendix A).
[c] Estimate for 1959 linear and unreliable (see Appendix A).

115

TABLE 11A – CHANGES IN THE DRILLING COST FUNCTION: 1955-56 – 1959
Summary: Dry Wells

		Definite decrease	Probable decrease	Ambiguous or no change	Probable increase	Definite increase
K_1/K_0		11, 11542	2, 547	2, 2245	2, 1385	9, 3256
$\alpha_1 - \alpha_0$		10, 4093	2, 684	1, 1937	4, 3787	9, 8479
H_1/H_0		14, 12655	1, 1937	0	3, 1817	8, 2566
MC_1/MC_0:	5,000 ft.	17, 13209	2, 2438	3, 906	0	4, 2422
	10,000 ft.	13, 5461	3, 3548	1, 1937	3, 2116	6, 5913
	15,000 ft.	11, 4211	0	2, 3684	7, 5158	6, 5922
Y_1/Y_0:	5,000 ft.	15, 14423	0	3, 1003	1, 158	7, 3391
	10,000 ft.	16, 8251	2, 4490	5, 4372	1, 308	2, 1554
	15,000 ft.	13, 5461	3, 3548	1, 1937	3, 2107	6, 5922

Grand Total

26, 18975

TABLE 11B – CHANGES IN THE DRILLING COST FUNCTION: 1955-56 – 1959
Summary: Productive Wells

		Definite decrease	Probable decrease	Ambiguous or no change	Probable increase	Definite increase
K_1/K_0		9, 10392	1, 3175	2, 798	6, 5863	8, 10120
$\alpha_1 - \alpha_0$		9, 10443	2, 3996	2, 726	4, 2064	9, 13119
H_1/H_0		9, 9623	1, 3175	3, 5014	3, 1820	10, 10716
MC_1/MC_0:	5,000 ft.	11, 11377	2, 2238	0	2, 1341	11, 15392
	10,000 ft.	7, 8085	5, 5552	0	4, 1778	10, 14933
	15,000 ft.	7, 8085	3, 4291	1, 448	5, 3841	10, 13683
Y_1/Y_0:	5,000 ft.	13, 16209	1, 2053	2, 4193	1, 323	9, 7570
	10,000 ft.	10, 10408	4, 3711	1, 1093	1, 1018	10, 14118
	15,000 ft.	7, 8085	4, 4739	1, 813	4, 1778	10, 14933

Grand Total

26, 30348

Appendices to Study II

Appendix A
Estimating Procedures and Measures of Goodness of Fit

The function (9) is not linear in the parameters. In particular, it is not linear in α, although it is in K. Moreover, no transformation of the variables which does not involve one or both of the parameters will make it simultaneously linear in the two parameters. It is therefore necessary to fit it by iterative techniques.

This was done in the following manner. The sum of squares of residuals (the differences between actual values of Y and values predicted from equation (9)[1]) weighted by the number of wells in each depth class was expressed as a function of the two parameters to be estimated. That function was differentiated and the derivatives set equal to zero, the result being two non-linear minimizing equations. The least squares estimates of the parameters were then approximated by expanding the first derivatives of the sum-of-squares function in Taylor Series in the parameters about an initial guess, dropping all non-linear terms and solving the resulting equations for the differences between the initial guesses and the least-squares parameter estimates. These differences were then added to the initial guesses to obtain a new set of guesses and the procedure repeated until both parameters failed to change by more than .001 per cent.[2]

When convergence had been obtained, the matrix of second derivatives of the function to be minimized was inverted and multiplied by the estimated residual variance to secure the asymptotic variance-covariance matrix of α and K.[3] (Small sample significance tests are unknown here.) These were then used to compute the asymptotic standard error of H.[4]

These asymptotic standard errors are one measure of goodness of fit; however, other, over-all measures are also desirable. These are provided

---

[1] While it might be more plausible to assume a multiplicative rather than an additive residual, the fit is so good as to make little difference to the estimates. This small practical difference was shown by considerable experimentation which also showed that convergence was considerably easier to obtain with an additive residual. The difference in results thus did not appear great enough to justify the large added complication involved.

[2] This procedure is essentially that of the "method of scoring." See J. S. Cramer, *The Ownership of Major Consumer Durables* (Cambridge: Cambridge University Press, 1962), pp. 46–49.

[3] The matrix in question (when multiplied by the residual variance) is the negative of the matrix of second derivatives of the logarithm of the likelihood function, on the assumption that the residuals are normally distributed, and thus gives the desired asymptotic results. (See L. R. Klein, *A Textbook of Econometrics* [Evanston: Row, Peterson and Co., 1953], p. 55.) Since the assumption of normality is rather dubious, the asymptotic standard errors presented should be taken as merely indicative of goodness of fit. In practice, the matrix involved has already been inverted at the last step of the iteration.

[4] See Klein, *op. cit.*, p. 258.

by the (weighted) standard error of estimate and by the squared coefficient of determination, both uncorrected and corrected for degrees of freedom. The squared coefficient of determination is defined as one minus the ratio of the (weighted) sum of squares of residuals to the (weighted) sum of squares of the deviations of the dependent variable (Y) from its (weighted) mean. It is comparable as a measure of goodness of fit to the squared multiple correlation coefficient in linear regressions (which is identical with it in such cases). By convention, should the computed squared coefficient of determination be negative, which can happen in a non-linear problem, it is reported as zero.[5]

As has been indicated, the result of all this is to fit equation (9) to the data, weighting the observation for each depth class by the number of wells drilled therein[6] and weighting all goodness-of-fit measures similarly. This was done instead of the obvious, and slightly simpler, alternative of an unweighted regression, for the following reasons. First, in the case of the *JAS* sample, as has been indicated in Section 4, it is a waste of information to use only depth-class aggregates. Lacking access to the individual well data, the alternative is clearly to weight the observations by the number of wells reporting to take account of the fact that the figures for some depth classes are based on more information than those for others.[7] To the extent that reporting is proportional to the number of wells drilled over depth classes —which is not wholly the case, as indicated on page 59—a similar argument leads to the weighting of depth class observations by number of wells when using the *JAS* figures as published.

In addition, however, there are two reasons for applying this procedure to the *JAS* figures as published. The first is to obtain comparability of treatment and hence to facilitate comparison of our results from the two sets of data; more important, however, is the fact that one ought plausibly to want a cost function which does well in depths where a large number of wells are drilled.[8]

What are the results of the weighting procedure? It should not be thought

[5] See F. C. Mills, *Statistical Methods* (3rd ed.; New York: Henry Holt and Co., 1955), pp. 584–88. (Mills uses the term "index of correlation" here instead of the more general term. See *ibid.*, pp. 645–46.)

[6] That is, the number of wells in that depth class as shown in the data involved. This is the total number of wells drilled in the case of the *JAS* figures as published and the number of wells reporting in the case of the *JAS* sample.

[7] If the individual well residual variance is the same for all wells, the residual variance for a depth class mean is inversely proportional to the number of wells reporting in that depth class. We are thus weighting inversely to variance. See Klein, *op. cit.*, pp. 309–11.

[8] The obvious alternative to this is to secure a cost function which does relatively better at high depths. This is largely taken care of automatically, however, by the fact that the observations for very deep wells are extreme ones and therefore receive high weight in view of their contribution to over-all variance. Since, as discussed below, an excellent fit to the unweighted data is generally obtained, the difference in the weights applied to the sample and published data has small effect on our comparisons thereof. The sample data must be weighted by number of wells reporting and this cannot be done in the case of the published data, where figures for depth-classes not in the sample are frequently given.

that it biases upward the measures of goodness of fit in cases where the distribution of wells over depth classes is very uneven. Even if nearly all drilling occurs in one depth class, for example, so that the weighted residual variance about any function which passes close to the point of means for that depth class will be very small, the squared coefficient of determination will not automatically be near unity. This is so because the weighted variance of Y around its own weighted mean will be similarly small, and it is the ratio of the two which is subtracted from unity in computing the coefficient. It thus remains true that to obtain a high squared coefficient of determination, it is necessary for the function as fitted to do substantially better in explaining the dependent variable than an estimate which places that variable at its mean. However, the weighted standard error of estimate will be low in such cases, as will the asymptotic standard errors of the coefficients.

In view of all this, it seems clearly desirable to compute an unweighted measure of goodness of fit to see how well our estimated function comes in each case to fitting the depth class means without adjustment for number of wells. Accordingly, an unweighted squared coefficient of determination was computed. Even this measure is usually rather high in the results, although it is substantially below the weighted measure in some cases, as is to be expected, since the unweighted measure is not being maximized by our parameter estimates. (It is interesting to observe, incidentally, that asymptotic standard errors are usually relatively high when the unweighted squared coefficient of determination is relatively low.) The true test of the power of our model is the weighted coefficient; the fact that an unweighted measure is also frequently large is a reassuring bonus.

There are two other matters which need brief discussion. First, the estimating procedure outlined above requires an initial parameter estimate sufficiently close to the least squares estimates to allow the iterative procedure to converge. This is by no means a trivial matter.[9] In general, where this proved to be a difficulty, the following systematic procedure was adopted.

Since equation (9) is linear in K, it is easy to minimize the residual sum of squares with respect to K, given a value of α. Accordingly, for each data case, various values of α were chosen and the conditionally minimizing value of K was computed for each one as well as the corresponding value of the weighted squared coefficient of determination.[10] It was thus possible to

---

[9] Words cannot convey the extent to which this is true. Indeed, the experience of this study leads me to believe that, however straightforward in principle, non-linear estimation is prohibitive in practice, even with modern computing equipment, unless the number of essential non-linearities is low (it is 1 in this case) or the number of data cases are not excessive (approximately 400 here).

[10] This was used rather than the sum of squares to be minimized for reasons of ease of scanning. Maximizing the squared coefficient of determination is obviously equivalent to minimizing the residual sum of squares. For a more detailed description of this general procedure, see M. Nerlove, *The Dynamics of Supply: Estimation of Farmers' Response to Price* (Baltimore: Johns Hopkins Press, 1958), pp. 187–89.

obtain, for each data case, the latter coefficient as a function of the choice of α and such choices could be spaced as narrowly as desired. For our initial estimates, then, that value of α was chosen for which the function had its maximum, along with the corresponding value of K.

This procedure worked extremely well in nearly all cases and also provided a check on whether the maximum attained by the iterative procedure was in fact the *maximum maximorum*.[11] In most cases there was no problem of this kind, there being only one maximum. In about 5 per cent of the data cases, however, such a problem did arise; indeed, in these cases there was a very flat plateau as α approached zero, with occasional very minor hills and valleys. In such cases, it proved impossible to attain convergence by the methods described. This brings us to the final matter in this appendix.

As is evident in the results, in nearly all cases, α as estimated is roughly on the order of magnitude of 10^{-4}. This corresponds to an increase in marginal cost of about 10 per cent every 1,000 feet.[12] In the cases now under discussion, however, the plateau occurred with α on the order of magnitude of 10^{-8} (roughly). This corresponds to an increase in marginal cost of about one-hundredth of one per cent in 10,000 feet. It is thus apparent that these are cases in which the function is only negligibly different over the relevant range from a straight line through the origin.[13] Accordingly, in all such cases we fitted such a straight line. It is reassuring to note, however, that the data areas in which this phenomenon occurs are generally not ones with a great amount of drilling. In general, the difficulty arises when cost per well appears to *decline* with depth over some ranges (presumably due to inadequate reporting or to geological heterogeneity). It is clearly a negligible problem.[14]

---

[11] A check for a *relative* maximum was built in by checking the positive definiteness of the matrix of second derivatives (positive definite because the residual sum of squares was being directly minimized).

[12] Y is in dollars and X in feet throughout.

[13] See footnote 9 of Study II.

[14] A minor question arises as to the number of degrees of freedom in such straight-line cases. Here only one parameter is involved, since the line must pass through the origin. However, a two-parameter analysis was required to discover this. Degrees of freedom have been reported on the basis of only one parameter, and no attempt has been made to find similar cases by searching the data areas for which only two depth classes are given.

Appendix B
The Drilling Cost Results in Detail

Figures in parentheses are *asymptotic* standard errors save in the cases discussed at the end of Appendix A (which will be so indicated) where they are simply standard errors. Aside from notation already employed, the following symbols are used:

I^2 = weighted squared coefficient of determination.

$\bar{I}^2 = I^2$ corrected for degrees of freedom.

J^2 = unweighted squared coefficient of determination.

S = standard error of estimate (weighted).

In all the following, X is in feet and Y in current dollars (except where indicated).

Table B1 – Drilling Cost Regressions: Joint Association Survey Sample, 1959

Data area	Type of well	K ($\times 10^4$ omitted)	α ($\times 10^{-4}$ omitted)	H (current dollars)	I^2	J^2	$\bar{I}^2$	S (current dollars)	Wells	Degrees of freedom
Alabama	Dry	7.176 (3.968)	0.8170[a] (0.2909)	5.863[b] (1.170)	.9956	.9945	.9867	4,380	4	1
	Productive	Insufficient data								
	Oil	Insufficient data								
	Gas	No gas wells drilled								
Alaska	Dry[l]	∞	0	178.5[a] (38.23)	.2673	.2860	.2673	24,880	4	2
	Productive[l]	∞	0	63.66[a,p] (19.58)	.5419	.3019	.5419	230,100	6	3
	Oil	Insufficient data								
	Gas	Insufficient data								
Appalachian[t]	Dry	1.628[a] (0.3942)	2.967[c] (0.2630)	4.830[b] (0.7531)	.9834	.9597	.9751	9,628	37	4
	Productive	3.790[d] (0.09093)	1.807[e] (0.02563)	6.849[e] (0.06867)	.9998	.9966	.9998	358.7	198	4
	Oil	0.3145[c] (0.02828)	6.931[d] (0.2905)	2.180[d] (0.1065)	.9995	.9965	.9985	245.2	21	1
	Gas	4.186[d] (0.1341)	1.706[e] (0.03318)	7.143[e] (0.09183)	.9990	.9944	.9984	433.2	177	3
Arkansas	Dry	2.929 (1.551)	1.557[a] (0.4840)	4.559[a] (1.010)	.9803	.7021	.9671	5,140	30	3
	Productive	3.968[b] (0.7522)	2.102[c] (0.1866)	8.343[b] (0.8635)	.9682	.9697	.9555	7,281	39	5
	Oil	1.680[b] (0.2799)	2.834[c] (0.2164)	4.762[c] (0.4422)	.9906	.9784	.9859	2,878	28	4
	Gas	15.34 (9.098)	1.049[a] (0.4136)	16.09[a] (3.295)	.9511	.9097	.9184	19,130	11	3

Table B1 — Drilling Cost Regressions: Joint Association Survey Sample, 1959 (continued)

Data area	Type of well	K ($\times 10^4$ omitted)	α ($\times 10^{-4}$ omitted)	H (current dollars)	r^2	J^2	$\bar{r}^2$	S (current dollars)	Wells	Degrees of freedom
Onshore California	Dry	9.918[b] (1.446)	0.9645[c] (0.07951)	9.565[c] (0.6225)	.9670	.7724	.9575	9,475	166	7
	Productive	14.47[d] (0.3689)	0.9327[e] (0.01504)	13.49[e] (0.1294)	.9970	.9902	.9960	928.0	723	6
	Oil	15.32[d] (0.4204)	0.9034[e] (0.01584)	13.84[e] (0.1401)	.9970	.9896	.9961	979.0	687	6
	Gas	22.69[a] (7.132)	0.5763[a] (0.1451)	13.08[c] (0.8489)	.9942	.9541	.9918	4,964	36	5
Offshore California	Dry	Insufficient data								
	Productive	5.302[b] (1.038)	2.754[c] (0.1891)	14.60[b] (1.882)	.9800	.9806	.9700	20,330	27	4
	Oil	5.302[b] (1.038)	2.754[c] (0.1891)	14.60[b] (1.882)	.9800	.9806	.9700	20,330	27	4
	Gas	No gas wells drilled								
Total California	Dry	10.40[b] (1.495)	0.9404[c] (0.07761)	9.778[c] (0.6158)	.9694	.7824	.9607	9,178	168	7
	Productive	15.78[d] (0.4578)	0.8956[e] (0.01675)	14.13[e] (0.1492)	.9964	.9864	.9952	1,063	750	6
	Oil	16.85[d] (0.5241)	0.8627[e] (0.01752)	14.54[e] (0.1607)	.9964	.9855	.9953	1,116	714	6
	Gas	22.69[a] (7.132)	0.5763[a] (0.1451)	13.08[c] (0.8489)	.9942	.9541	.9918	4,964	36	5
Colorado	Dry	0.5580[c] (0.04443)	3.947[d] (0.09007)	2.202[c] (0.1264)	.9887	.9780	.9841	3,115	159	5
	Productive	1.574[b] (0.2248)	3.094[d] (0.1634)	4.870[c] (0.4442)	.9919	.9401	.9887	6,654	100	5
	Oil	3.366 (3.026)	1.819 (0.9956)	6.123[a] (2.182)	.9710	0	.9516	10,140	61	3
	Gas	5.620[b] (0.7298)	1.900[c] (0.1240)	10.68[c] (0.6984)	.9955	.9793	.9936	5,977	39	5

Data area	Type of well	K ($\times 10^4$ omitted)	α ($\times 10^{-4}$ omitted)	H (current dollars)	I^2	J^2	$\bar{I}^2$	S (current dollars)	Wells	Degrees of freedom
Illinois	Dry	2.565[a] (0.9193)	1.541[a] (0.4476)	3.954[c] (0.2725)	.9984	.9768	.9951	264.4	34	1
	Productive	3.337[c] (0.2744)	2.089[c] (0.1238)	6.970[d] (0.1740)	.9976	.9919	.9959	518.3	237	3
	Oil	3.337[c] (0.2744)	2.089[c] (0.1238)	6.970[d] (0.1740)	.9976	.9919	.9959	518.3	237	3
	Gas	No data in sample[j]								
Indiana	Dry[l]	∞	0	3.519[c,r] (0.2987)	.9093	.9384	.9093	865.4	13	2
	Productive[l]	∞	0	10.40[f,s] (0.08499)	.9986	.9972	.9986	383.0	26	2
	Oil[l]	∞	0	10.40[f,s] (0.08499)	.9986	.9972	.9986	383.0	26	2
	Gas	No data in sample[j]								
Kansas	Dry	0.4312[d] (0.02031)	4.217[e] (0.08115)	1.818[d] (0.05107)	.9928	.9715	.9880	308.2	399	3
	Productive	0.5126[c] (0.03879)	4.809[d] (0.1390)	2.465[d] (0.1161)	.9776	.8985	.9626	667.8	589	3
	Oil	2.242[c] (0.1844)	2.414[c] (0.1232)	5.414[d] (0.1712)	.9934	.9494	.9890	433.6	507	3
	Gas	2.266[a] (0.8162)	2.430[a] (0.4874)	5.504[b] (0.8832)	.9928	.7838	.9783	203.6	82	1
Kentucky	Dry	1.265[a] (0.5065)	2.949[a] (0.8316)	3.729[b] (0.4477)	.9968	.9167	.9902	498.3	48	1
	Productive	8.093[b] (1.012)	0.9825[b] (0.1043)	7.952[e] (0.1537)	.9983	.9904	.9966	220.7	145	2
	Oil[l]	∞	0	8.736[d,r] (0.3823)	.9815	.9788	.9815	967.3	75	2
	Gas	6.044[b] (1.203)	1.251[b] (0.1996)	7.560[d] (0.3023)	.9989	.9787	.9978	450.1	70	2

Data area	Type of well	K ($\times 10^4$ omitted)	α ($\times 10^{-4}$ omitted)	H (current dollars)	I^2	J^2	$\bar{I}^2$	S (current dollars)	Wells	Degrees of freedom
North Louisiana	Dry	0.8881^c (0.04813)	2.750^e (0.05076)	2.442^d (0.08784)	.9987	.9957	.9980	1,350	48	4
	Productive	0.5191^b (0.07781)	3.873^d (0.1482)	2.011^b (0.2249)	.9868	.9801	.9803	3,455	132	4
	Oil	2.789^b (0.3584)	1.881^c (0.1202)	5.246^c (0.3484)	.9948	.9839	.9922	2,363	110	4
	Gas	0.2549^a (0.1229)	4.667^b (0.4751)	1.190^a (0.4530)	.9970	.9423	.9911	13,280	22	1
South Louisiana	Dry	1.053^d (0.05138)	2.467^e (0.03338)	2.597^d (0.09191)	.9986	.9746	.9982	3,795	309	6
	Productive	3.102^d (0.06700)	1.927^f (0.01455)	5.976^e (0.08446)	.9990	.9897	.9987	2,286	540	7
	Oil	4.276^d (0.1443)	1.671^e (0.02278)	7.143^d (0.1450)	.9969	.9861	.9960	3,090	347	7
	Gas	2.535^c (0.1332)	2.089^e (0.03502)	5.294^d (0.1902)	.9964	.9798	.9939	4,703	193	3
Onshore Louisiana	Dry	1.085^d (0.02809)	$2 448^f$ (0.01776)	2.655^e (0.04966)	.9994	.9925	.9992	1,951	357	6
	Productive	3.227^e (0.03793)	1.903^g (0.007934)	6.141^f (0.04687)	.9996	.9967	.9995	1,172	672	7
	Oil	4.064^d (0.09969)	1.703^f (0.01668)	6.921^e (0.1030)	.9975	.9921	.9968	2,055	457	7
	Gas	2.949^d (0.1023)	1.994^e (0.02311)	5.881^d (0.1365)	.9981	.9901	.9969	3,111	215	3
Offshore Louisiana	Dry	7.862^b (1.301)	1.520^c (0.1086)	11.95^c (1.132)	.9901	.8923	.9862	19,700	88	5
	Productive	6.491^c (0.6010)	1.824^d (0.06595)	11.84^c (0.6753)	.9869	.8851	.9782	13,240	300	3
	Oil	7.768^b (1.045)	1.647^c (0.09910)	12.80^c (0.9634)	.9931	.8361	.9884	14,040	247	3
	Gas	10.11^a (2.992)	1.596^b (0.1889)	16.14^b (2.883)	.9614	.6999	.9356	49,140	53	3

Table B1 – Drilling Cost Regressions: Joint Association Survey Sample, 1959 (continued)

Data area	Type of well	K ($\times 10^4$ omitted)	α ($\times 10^{-4}$ omitted)	H (current dollars)	I^2	J^2	$\bar{I}^2$	S (current dollars)	Wells	Degrees of freedom
Total Louisiana	Dry	2.087^c (0.1406)	2.094^d (0.04594)	4.371^d (0.1995)	.9954	.9222	.9940	6,053	445	7
	Productive	6.572^e (0.06600)	1.565^f (0.006655)	10.28^f (0.06013)	.9991	.9955	.9988	1,230	972	7
	Oil	9.189^d (0.2012)	1.310^e (0.01429)	12.04^e (0.1343)	.9940	.9879	.9923	2,150	704	7
	Gas	4.714^d (0.2334)	1.784^e (0.03245)	8.412^d (0.2649)	.9946	.9706	.9910	5,070	268	3
Michigan	Dry	18.36 (91.71)	0.3675 (1.719)	6.748^a (2.160)	.9906	.4821	.9717	2,325	21	1
	Productive	4.070 (6.425)	2.248 (2.337)	9.151 (4.986)	.9828	0	.9485	9,311	58	1
	Oil	3.579 (5.626)	2.436 (2.436)	8.718 (5.037)	.9803	0	.9409	9,809	56	1
	Gas	Insufficient data								
Mississippi	Dry	1.588^a (0.5404)	1.902^b (0.2425)	3.021^a (0.6474)	.9669	0	.9503	12,920	94	4
	Productive	0.7346^a (0.1532)	2.819^c (0.1559)	2.071^b (0.3183)	.9723	.7790	.9584	12,230	144	4
	Oil	0.5384^a (0.1406)	3.069^c (0.1986)	1.652^b (0.3255)	.9768	.7743	.9651	14,760	133	4
	Gas	3.687 (2.123)	1.681^a (0.3726)	6.197^a (2.203)	.9624	.7987	.9374	31,580	11	3
Montana	Dry	1.150^a (0.4887)	2.864^b (0.3724)	3.292^a (0.9770)	.9423	.9321	.9192	19,120	47	5
	Productive	28.57^a ˙ (12.48)	0.4189^a (0.1548)	11.97^c (0.8177)	.9986	.9069	.9976	3,038	67	3
	Oil	20.21^b (3.946)	0.5558^e (0.008698)	11.23^d (0.4418)	.9997	.9761	.9995	1,556	64	2
	Gas	Insufficient data								

126

Data area	Type of well	K ($\times 10^4$ omitted)	α ($\times 10^{-4}$ omitted)	H (current dollars)	I^2	J^2	$\bar{I}^2$	S (current dollars)	Wells	Degrees of freedom
Nebraska	Dry[l]	∞	0	3.684[d,s] (0.08595)	.8268	0	.8268	3,545	114	4
	Productive	Insufficient data								
	Oil	Insufficient data								
	Gas	Insufficient data								
Northwest New Mexico	Dry	22.04 (12.34)	0.5944[a] (0.2476)	13.10[b] (2.020)	.8970	.8458	.8558	16,980	27	5
	Productive	6.230[a] (1.862)	1.524[b] (0.2841)	9.496[b] (1.102)	.9591	0	.9427	7,227	374	5
	Oil	56.67 (124.65)	0.2180 (0.4499)	12.36[b] (1.749)	.9872	0	.9821	6,066	225	5
	Gas	2.892[d] (0.09593)	2.510[e] (0.03841)	7.257[e] (0.1311)	.9987	.9940	.9980	967.1	149	4
Southeast New Mexico	Dry	11.16[b] (1.192)	0.9239[c] (0.06136)	10.31[d] (0.4246)	.9906	.9505	.9875	3,901	95	6
	Productive	10.22[c] (0.8562)	0.9804[c] (0.05314)	10.02[d] (0.3095)	.9844	.9483	.9792	2,784	433	6
	Oil	10.06[b] (1.034)	0.9821[c] (0.06479)	9.878[d] (0.3790)	.9776	.9316	.9702	3,395	389	6
	Gas	5.629[a] (1.663)	1.517[b] (0.2488)	8.539[b] (1.148)	.9808	.9166	.9712	9,390	44	4
Total New Mexico	Dry	13.04[b] (1.368)	0.8391[c] (0.05721)	10.94[d] (0.4118)	.9887	.9437	.9850	3,697	122	6
	Productive	11.79[c] (0.9115)	0.9213[c] (0.04917)	10.86[d] (0.2750)	.9767	.9322	.9689	2,351	807	6
	Oil	10.32[c] (0.9328)	0.9620[c] (0.05792)	9.929[d] (0.3154)	.9812	.9340	.9749	2,698	614	6
	Gas	5.156[b] (0.5965)	1.784[c] (0.1127)	9.198[c] (0.4954)	.9928	.9241	.9899	4,283	193	5

Table B1 — Drilling Cost Regressions: Joint Association Survey Sample, 1959 (continued)

Data area	Type of well	K ($\times 10^4$ omitted)	α ($\times 10^{-4}$ omitted)	H (current dollars)	I^2	J^2	I^2	S (current dollars)	Wells	Degrees of freedom
North Dakota	Dry	1.702[c] (0.09675)	2.152[d] (0.05147)	3.663[d] (0.1220)	.9980	.9894	.9970	1,194	95	4
	Productive	1.712[c] (0.1352)	2.418[d] (0.07065)	4.141[c] (0.2071)	.9930	.9788	.9884	2,402	137	3
	Oil	1.740[c] (0.1395)	2.402[d] (0.07199)	4.179[c] (0.2110)	.9928	.9784	.9880	2,405	136	3
	Gas	Insufficient data								
Oklahoma	Dry	1.634[d] (0.03884)	2.351[f] (0.01512)	3.842[e] (0.06729)	.9978	.9978	.9972	2,018	382	7
	Productive	2.591[d] (0.1243)	2.231[e] (0.03933)	5.779[d] (0.1801)	.9932	.9674	.9913	3,818	1,053	7
	Oil	2.431[a] (0.6274)	2.232[b] (0.2494)	5.425[b] (0.8048)	.9898	.7980	.9864	8,469	795	6
	Gas	3.318[c] (0.2161)	2.081[d] (0.04823)	6.906[d] (0.2967)	.9876	.9645	.9841	7,889	258	7
East Texas	Dry	0.7909[c] (0.07498)	2.839[d] (0.08354)	2.245[c] (0.1482)	.9897	.9886	.9856	3,580	64	5
	Productive	0.9884[c] (0.09317)	2.974[d] (0.08827)	2.939[c] (0.1920)	.9882	.9532	.9842	5,009	181	6
	Oil	0.5804[b] (0.06727)	3.556[d] (0.1233)	2.064[c] (0.1688)	.9765	.9786	.9672	3,601	142	5
	Gas	1.333[d] (0.05514)	2.701[e] (0.03523)	3.600[d] (0.1029)	.9998	.9966	.9996	2,367	39	3
Gulf Coast Texas	Dry	0.4407[d] (0.02092)	3.382[e] (0.03896)	1.491[d] (0.05381)	.9968	.9923	.9958	2,599	176	6
	Productive	1.556[e] (0.02567)	2.548[e] (0.01241)	3.966[e] (0.04657)	.9983	.9979	.9979	1,644	368	7
	Oil	1.039[c] (0.06424)	2.928[e] (0.05560)	3.042[d] (0.1313)	.9893	.9884	.9857	3,223	235	6
	Gas	1.698[d] (0.04406)	2.482[f] (0.01856)	4.216[e] (0.07839)	.9990	.9974	.9987	2,956	133	6

Data area	Type of well	K ($\times 10^4$ omitted)	α ($\times 10^{-4}$ omitted)	H (current dollars)	I^2	J^2	$\bar{I}^2$	S (current dollars)	Wells	Degrees of freedom
North Central Texas	Dry	1.210^a (0.2520)	2.507^c (0.2030)	3.033^b (0.3990)	.9810	.7223	.9746	5,662	220	6
	Productive	2.180^a (0.4499)	2.250^b (0.2712)	4.904^c (0.4281)	.9941	.9146	.9912	1,752	518	4
	Oil	2.138^a (0.5033)	2.253^b (0.3120)	4.817^c (0.4750)	.9925	.9042	.9888	1,937	480	4
	Gas	6.237^a (1.339)	1.220^b (0.1883)	7.610^c (0.4639)	.9993	.9789	.9985	1,065	38	2
Panhandle Texas	Dry	10.31 (17.50)	0.7830 (0.9892)	8.071^a (3.558)	.9425	.5495	.9195	15,910	104	5
	Productive	6.424^b (0.6492)	1.282^c (0.08226)	8.237^d (0.3100)	.9971	.9531	.9956	1,656	388	4
	Oil	5.253^b (0.5760)	1.450^c (0.09622)	7.619^d (0.3371)	.9968	.9550	.9952	1,968	248	4
	Gas	8.289^a (3.188)	1.096^a (0.2815)	9.087^b (1.180)	.9936	.8908	.9905	5,479	140	4
Southwest Texas	Dry	0.9711^d (0.04826)	2.731^e (0.03705)	2.652^d (0.09659)	.9976	.9952	.9969	3,165	186	7
	Productive	3.819^c (0.3075)	1.825^d (0.06483)	6.969^d (0.3212)	.9923	.9396	.9901	5,132	428	7
	Oil	1.502^a (0.3993)	2.714^b (0.2847)	4.075^b (0.6608)	.9631	.8753	.9483	5,332	281	5
	Gas	5.086^c (0.4461)	1.621^d (0.06596)	8.242^d (0.3953)	.9961	.9592	.9950	7,117	147	7
West Texas	Dry	1.577^d (0.04650)	2.303^f (0.01722)	3.631^d (0.08046)	.9941	.9970	.9924	3,036	345	7
	Productive	5.835^d (0.1646)	1.416^e (0.02152)	8.263^e (0.1106)	.9988	.9775	.9984	1,267	1,938	7
	Oil	6.364^f (0.04576)	1.333^g (0.005414)	8.484^g (0.02738)	.9988	.9965	.9984	280.1	1,878	6
	Gas	5.161^b (0.9759)	1.638^c (0.1354)	8.451^b (0.9102)	.9941	.8535	.9918	13,920	60	5

Data area	Type of well	K $(\times 10^4$ omitted)	α $(\times 10^{-4}$ omitted)	H (current dollars)	I^2	J^2	$\bar{I}^2$	S (current dollars)	Wells	Degrees of freedom
Offshore Texas	Dry	Insufficient data								
	Productive	Insufficient data								
	Oil	Insufficient data								
	Gas	Insufficient data								
Total Texas	Dry	1.433[f] (0.08078)	2.376[g] (0.003623)	3.405[g] (0.01418)	.9986	.9970	.9982	519.8	1,096	7
	Productive	2.975[d] (0.08591)	1.995[e] (0.02383)	5.935[e] (0.1025)	.9956	.9723	.9943	1,633	3,823	7
	Oil	4.227[f] (0.03538)	1.647[g] (0.006980)	6.963[g] (0.02956)	.9981	.9931	.9975	325.0	3,265	6
	Gas	2.679[f] (0.02580)	2.141[g] (0.007298)	5.735[f] (0.03616)	.9986	.9986	.9982	919.5	558	7
Utah	Dry	7.263[a] (2.062)	1.448[b] (0.2387)	10.52[b] (1.302)	.9762	.8862	.9642	12,980	62	4
	Productive	4.147[a] (1.685)	2.179[b] (0.3963)	9.036[a] (2.137)	.9654	.6685	.9308	27,170	117	2
	Oil	4.147 (2.190)	2.179[a] (0.5142)	9.035[a] (2.781)	.9517	.3667	.9034	35,430	115	2
	Gas	Insufficient data								
Wyoming	Dry	1.432[b] (0.1444)	2.629[d] (0.08830)	3.765[c] (0.2567)	.9722	.9648	.9630	5,787	175	6
	Productive	4.896[c] (0.2859)	1.694[d] (0.04894)	8.296[d] (0.2571)	.9640	.9895	.9521	4,001	283	6
	Oil	154.6 (942.0)	0.08871 (0.5248)	13.71[b] (2.470)	.9330	.6361	.9061	6,772	238	5
	Gas	9.696[a] (2.297)	1.358[b] (0.1624)	13.16[b] (1.616)	.9800	.9061	.9720	26,710	45	5

Footnotes

[a]Coefficient greater than 2 times asymptotic standard error.
[b]Coefficient greater than 5 times asymptotic standard error.
[c]Coefficient greater than 10 times asymptotic standard error.
[d]Coefficient greater than 20 times asymptotic standard error.
[e]Coefficient greater than 50 times asymptotic standard error.
[f]Coefficient greater than 100 times asymptotic standard error.
[g]Coefficient greater than 200 times asymptotic standard error.
[j]But JAS reports figures for one depth class.
[k]Sum of North and South Louisiana data. Computed for reasons of later comparability.
[l]Linear regression performed; see Appendix A, pp. 117-20. In these cases $\alpha = 0$, $K = \infty$, and the coefficient of the regression is given as H. I^2 and $\bar{I}^2$
  are identical by definition (since the number of degrees of freedom lost in both numerator and denominator of the relevant fraction is one).
[p]Coefficient significant at 5% level.
[q]Coefficient significant at 2% level.
[r]Coefficient significant at 1% level.
[s]Coefficient significant at 1/10% level.
[t]Includes New York, Ohio, Pennsylvania, and West Virginia.

N.B. These significance levels should be interpreted with care. The slope of a straight line forced through the origin will be significantly different
     from zero roughly so long as the point of mean of the observations lies away from the axes, regardless of the goodness of fit of the line.

Data area	Type of well	α ($\times 10^{-4}$ omitted)	Depth increase (feet to nearest hundred) for which costs double at initial depth of				Limit = depth increase for which marginal cost doubles
			0 ft.	5,000 ft.	10,000 ft.	15,000 ft.	
Alabama	Dry	0.8170 (0.2909)	0	3,500	5,400	6,500	8,500
	Productive	Insufficient data					
	Oil	Insufficient data					
	Gas	Insufficient data					
Alaska	Dry	0	0	5,000	10,000	15,000	∞
	Productive	0	0	5,000	10,000	15,000	∞
	Oil	Insufficient data					
	Gas	Insufficient data					
Appalachian	Dry	2.967 (0.2630)	0	1,900	2,200	2,300	2,300
	Productive	1.807 (0.02563)	0	2,600	3,400	3,600	3,800
	Oil	6.931 (0.2905)	0	1,000	1,000	1,000	1,000
	Gas	1.706 (0.03318)	0	2,700	3,500	3,800	4,100
Arkansas	Dry	1.557 (0.4840)	0	2,800	3,700	4,100	4,500
	Productive	2.102 (0.1866)	0	2,400	3,000	3,200	3,300
	Oil	2.834 (0.2164)	0	2,000	2,300	2,400	2,400
	Gas	1.049 (0.4136)	0	3,300	5,100	5,600	6,600
Onshore California	Dry	0.9645 (0.07951)	0	3,400	5,000	5,900	7,200
	Productive	0.9327 (0.01504)	0	3,400	5,100	6,000	7,400
	Oil	0.9034 (0.01584)	0	3,400	5,200	6,100	7,700
	Gas	0.5763 (0.1451)	0	3,900	6,300	7,900	12,000

Data area	Type of well	α ($\times 10^{-4}$ omitted)	Depth increase (feet to nearest hundred) for which costs double at initial depth of				Limit = depth increase for which marginal cost doubles
			0 ft.	5,000 ft.	10,000 ft.	15,000 ft.	
Offshore California	Dry	Insufficient data					
	Productive	2.754 (0.1891)	0	2,000	2,400	2,500	2,500
	Oil	2.754 (0.1891)	0	2,000	2,400	2,500	2,500
	Gas	No gas wells drilled					
Total California	Dry	0.9404 (0.07761)	0	3,400	5,100	6,000	7,400
	Productive	0.8956 (0.01675)	0	3,400	5,200	6,100	7,700
	Oil	0.8627 (0.01752)	0	3,500	5,300	6,300	8,000
	Gas	0.5763 (0.1451)	0	3,900	6,300	7,900	12,000
Colorado	Dry	3.947 (0.09007)	0	1,600	1,700	1,800	1,800
	Productive	3.094 (0.1634)	0	1,900	2,200	2,200	2,200
	Oil	1.819 (0.9956)	0	2,600	3,300	3,600	3,800
	Gas	1.900 (0.1240)	0	2,500	3,200	3,500	3,600
Illinois	Dry	1.541 (0.4476)	0	2,800	3,800	4,200	4,500
	Productive	2.089 (0.1238)	0	2,400	3,000	3,200	3,300
	Oil	2.089 (0.1238)	0	2,400	3,000	3,200	3,300
	Gas	No data in sample					
Indiana	Dry	0	0	5,000	10,000	15,000	∞
	Productive	0	0	5,000	10,000	15,000	∞
	Oil	0	0	5,000	10,000	15,000	∞
	Gas	No data in sample					

Table B2 — Depth Increase for Which Costs Double (Selected Initial Depths)
Joint Association Survey Sample, 1959 (continued)

Data area	Type of well	α ($\times 10^{-4}$ omitted)	Depth increase (feet to nearest hundred) for which costs double at initial depth of				Limit = depth increase for which marginal cost doubles
			0 ft.	5,000 ft.	10,000 ft.	15,000 ft.	
Kansas	Dry	4.217 (0.08115)	0	1,500	1,600	1,600	1,600
	Productive	4.809 (0.1390)	0	1,300	1,400	1,400	1,400
	Oil	2.414 (0.1232)	0	2,200	2,700	2,800	2,900
	Gas	2.430 (0.4874)	0	2,200	2,700	2,800	2,900
Kentucky	Dry	2.949 (0.8316)	0	1,900	2,300	2,400	2,400
	Productive	0.9825 (0.1043)	0	3,300	4,900	5,800	7,100
	Oil	0	0	5,000	10,000	15,000	∞
	Gas	1.251 (0.1996)	0	3,100	4,300	4,900	5,500
North Louisiana	Dry	2.750 (0.05076)	0	2,000	2,400	2,500	2,500
	Productive	3.873 (0.1482)	0	1,600	1,800	1,800	1,800
	Oil	1.881 (0.1202)	0	2,500	3,300	3,500	3,700
	Gas	4.667 (0.4751)	0	1,500	1,500	1,500	1,500
South Louisiana	Dry	2.467 (0.03338)	0	2,200	2,600	2,800	2,800
	Productive	1.927 (0.01455)	0	2,500	3,200	3,500	3,600
	Oil	1.671 (0.02278)	0	2,700	3,600	3,900	4,100
	Gas	2.089 (0.03502)	0	2,400	3,000	3,200	3,300
Onshore Louisiana	Dry	2.448 (0.01776)	0	2,200	2,600	2,800	2,800
	Productive	1.903 (0.007934)	0	2,500	3,200	3,500	3,600
	Oil	1.703 (0.01668)	0	2,700	3,600	3,900	4,100
	Gas	1.994 (0.02311)	0	2,500	3,100	3,400	3,500

Table B2 – Depth Increase for Which Costs Double (Selected Initial Depths)
Joint Association Survey Sample, 1959 (continued)

Data area	Type of well	a ($\times 10^{-4}$ omitted)	Depth increase (feet to nearest hundred) for which costs double at initial depth of				Limit = depth increase for which marginal cost doubles
			0 ft.	5,000 ft.	10,000 ft.	15,000 ft.	
Offshore Louisiana	Dry	1.520 (0.1086)	0	2,800	3,800	4,200	4,600
	Productive	1.824 (0.06595)	0	2,600	3,300	3,600	3,800
	Oil	1.647 (0.09910)	0	2,700	3,600	3,900	4,200
	Gas	1.596 (0.1889)	0	2,700	3,700	4,000	4,300
Total Louisiana	Dry	2.094 (0.04594)	0	2,400	3,000	3,200	3,300
	Productive	1.565 (0.006655)	0	2,800	3,700	4,100	4,400
	Oil	1.310 (0.01429)	0	3,000	4,200	4,700	5,300
	Gas	1.784 (0.03245)	0	2,600	3,400	3,700	3,900
Michigan	Dry	0.3675 (1.719)	0	4,200	7,300	9,600	18,900
	Productive	2.248 (2.337)	0	2,300	2,900	3,000	3,100
	Oil	2.436 (2.436)	0	2,200	2,700	2,800	2,800
	Gas	Insufficient data					
Mississippi	Dry	1.902 (0.2425)	0	2,500	3,200	3,500	3,600
	Productive	2.819 (0.1559)	0	2,000	2,400	2,500	2,500
	Oil	3.069 (0.1986)	0	1,900	2,200	2,300	2,300
	Gas	1.681 (0.3726)	0	2,700	3,600	3,900	4,100
Montana	Dry	2.864 (0.3724)	0	2,000	2,300	2,400	2,400
	Productive	0.4189 (0.1548)	0	4,100	7,000	9,100	16,500
	Oil	0.5558 (0.008698)	0	3,900	6,400	8,100	12,500
	Gas	Insufficient data					

135

Table B2 — Depth Increase for Which Costs Double (Selected Initial Depths)
Joint Association Survey Sample, 1959 (continued)

Data area	Type of well	α $(\times 10^{-4}$ omitted)	Depth increase (feet to nearest hundred) for which costs double at initial depth of				Limit = depth increase for which marginal cost doubles
			0 ft.	5,000 ft.	10,000 ft.	15,000 ft.	
Nebraska	Dry	0	0	5,000	10,000	15,000	∞
	Productive	Insufficient data					
	Oil	Insufficient data					
	Gas	Insufficient data					
Northwest New Mexico	Dry	0.5944 (0.2476)	0	3,800	6,200	7,800	11,700
	Productive	1.524 (0.2841)	0	2,800	3,800	4,200	4,500
	Oil	0.2180 (0.4499)	0	4,500	8,200	11,300	31,800
	Gas	2.510 (0.03841)	0	2,100	2,600	2,800	2,800
Southeast New Mexico	Dry	0.9239 (0.06136)	0	3,400	5,100	6,100	7,500
	Productive	0.9804 (0.05314)	0	3,300	4,900	5,800	7,100
	Oil	0.9821 (0.06479)	0	3,300	4,900	5,800	7,100
	Gas	1.517 (0.2488)	0	2,800	3,800	4,200	4,600
Total New Mexico	Dry	0.8391 (0.05721)	0	3,500	5,400	6,400	8,300
	Productive	0.9213 (0.04917)	0	3,400	5,100	6,100	7,500
	Oil	0.9620 (0.05792)	0	3,400	5,000	5,900	7,200
	Gas	1.784 (0.1127)	0	2,600	3,400	3,700	3,900
North Dakota	Dry	2.152 (0.05147)	0	2,400	2,900	3,100	3,200
	Productive	2.418 (0.07065)	0	2,200	2,700	2,800	2,900
	Oil	2.402 (0.07199)	0	2,200	2,700	2,800	2,900
	Gas	Insufficient data					

Table B2 — Depth Increase for Which Costs Double (Selected Initial Depths)
Joint Association Survey Sample, 1959 (continued)

Data area	Type of well	α ($\times 10^{-4}$ omitted)	Depth increase (feet to nearest hundred) for which costs double at initial depth of				Limit = depth increase for which marginal cost doubles
			0 ft.	5,000 ft.	10,000 ft.	15,000 ft.	
Oklahoma	Dry	2.351 (0.01512)	0	2,200	2,700	2,900	2,900
	Productive	2.231 (0.03933)	0	2,300	2,900	3,000	3,100
	Oil	2.232 (0.2494)	0	2,300	2,900	3,000	3,100
	Gas	2.081 (0.04823)	0	2,400	3,000	3,200	3,300
East Texas	Dry	2.839 (0.08354)	0	2,000	2,300	2,400	2,400
	Productive	2.974 (0.08827)	0	1,900	2,200	2,300	2,300
	Oil	3.556 (0.1233)	0	1,700	1,900	1,900	1,900
	Gas	2.701 (0.03523)	0	2,100	2,400	2,600	2,600
Gulf Coast Texas	Dry	3.382 (0.03896)	0	1,800	2,000	2,000	2,000
	Productive	2.548 (0.01241)	0	2,100	2,600	2,700	2,700
	Oil	2.928 (0.05560)	0	1,900	2,300	2,400	2,400
	Gas	2.482 (0.01856)	0	2,200	2,600	2,800	2,800
North Central Texas	Dry	2.507 (0.2030)	0	2,100	2,600	2,800	2,800
	Productive	2.250 (0.2712)	0	2,300	2,900	3,000	3,100
	Oil	2.253 (0.3120)	0	2,300	2,900	3,000	3,100
	Gas	1.220 (0.1883)	0	3,100	4,400	5,000	5,700
Panhandle Texas	Dry	0.7830 (0.9892)	0	3,600	5,500	6,700	8,900
	Productive	1.282 (0.08226)	0	3,000	4,200	4,800	5,400
	Oil	1.450 (0.09622)	0	2,900	3,900	4,400	4,800
	Gas	1.096 (0.2815)	0	3,200	4,700	5,400	6,300

Table B2 — Depth Increase for Which Costs Double (Selected Initial Depths)
Joint Association Survey Sample, 1959 (continued)

Data area	Type of well	a ($\times 10^{-4}$ omitted)	Depth increase (feet to nearest hundred) for which costs double at initial depth of				Limit = depth increase for which marginal cost doubles
			0 ft.	5,000 ft.	10,000 ft.	15,000 ft.	
Southwest Texas	Dry	2.731 (0.03705)	0	2,100	2,400	2,500	2,500
	Productive	1.825 (0.06483)	0	2,600	3,300	3,600	3,800
	Oil	2.714 (0.2847)	0	2,100	2,400	2,600	2,600
	Gas	1.621 (0.06596)	0	2,700	3,600	4,000	4,300
West Texas	Dry	2.303 (0.01722)	0	2,300	2,800	2,900	3,000
	Productive	1.416 (0.02152)	0	2,900	4,000	4,500	4,900
	Oil	1.333 (0.005414)	0	3,000	4,100	4,700	5,200
	Gas	1.638 (0.1354)	0	2,700	3,600	3,900	4,200
Offshore Texas	Dry	Insufficient data					
	Productive	Insufficient data					
	Oil	Insufficient data					
	Gas	Insufficient data					
Total Texas	Dry	2.376 (0.003623)	0	2,200	2,700	2,800	2,900
	Productive	1.995 (0.02383)	0	2,500	3,100	3,400	3,500
	Oil	1.647 (0.006980)	0	2,700	3,600	3,900	4,200
	Gas	2.141 (0.007298)	0	2,400	2,900	3,100	3,200
Utah	Dry	1.448 (0.2387)	0	2,900	3,900	4,400	4,800
	Productive	2.179 (0.3963)	0	2,400	2,900	3,100	3,200
	Oil	2.179 (0.5142)	0	2,400	2,900	3,100	3,200
	Gas	Insufficient data					

Table B2 — Depth Increase for Which Costs Double (Selected Initial Depths)
Joint Association Survey Sample, 1959 (continued)

Data area	Type of well	a ($\times 10^{-4}$ omitted)	Depth increase (feet to nearest hundred) for which costs double at initial depth of				Limit = depth increase for which marginal cost doubles
			0 ft.	5,000 ft.	10,000 ft.	15,000 ft.	
Wyoming	Dry	2.629 (0.08830)	0	2,100	2,500	2,600	2,600
	Productive	1.694 (0.04894)	0	2,700	3,600	3,900	4,100
	Oil	0.08871 (0.5248)	0	4,800	9,200	13,200	78,100
	Gas	1.358 (0.1624)	0	3,000	4,100	4,700	5,100

Table B3 — Drilling Cost Regressions: Joint Association Survey Data as Published, 1955, 1956, and 1959

Data area	Type of well	Year	K (× 10^4 omitted)	α (× 10^{-4} omitted)	H (current dollars per foot)	I^2	J^2	Î^2	S (current dollars)	Wells	Degrees of freedom
Alabama	Dry	1955	1.906^c (0.1317)	2.119^d (0.0678)	4.039^d (0.1542)	.9987	.9945	.9981	1,794	40	5
	Dry	1956	1.768^d (0.07711)	1.912^e (0.03192)	3.381^d (0.09232)	.9986	.9984	.9981	1,652	34	6
	Dry	1959	6.509^c (0.5966)	0.8642^f (0.05113)	5.625^d (0.1867)	.9975	.9950	.9966	1,516	25	6
	Productive	1955	Insufficient data								
	Productive	1956	6.007^b (1.193)	1.294^b (0.1360)	7.776^c (0.7280)	.9998	.9336	.9997	2,385	57	3
	Productive	1959	Insufficient data								
	Oil	1959	Insufficient data								
	Gas	1959	No gas wells drilled								
Alaska^j	Dry^o	1959	∞	0	191.2^{c,p} (15.77)	.7133	.6247	.7133	320,900	9	2
	Productive^o	1959	∞	0	69.39^{b,r} (9.803)	.5394	.5471	.5391	193,600	9	3
	Oil	1959	Insufficient data								
	Gas	1959	Insufficient data								

140

Table B3 – Drilling Cost Regressions: Joint Association Survey Data as Published, 1955, 1956, and 1955 (continued)

Data area	Type of well	Year	K ($\times 10^4$ omitted)	α ($\times 10^{-4}$ omitted)	H (current dollars per foot)	I^2	J^2	$\bar{I}^2$	S (current dollars)	Wells	Degrees of freedom
Appalachian[k]	Dry	1955	1.089^c (0.09013)	2.986^d (0.1170)	3.252^d (0.1506)	.9861	.9586	.9791	901.4	558	4
	Dry	1956	8.688 (8.426)	0.5744 (0.4884)	4.990^b (0.6701)	.8828	0	.8243	2,641	523	4
	Dry	1959	3.085^e (0.05524)	2.331^f (0.02025)	7.192^f (0.06976)	.9966	.9982	.9952	536.9	492	5
	Productive	1955	2.264^c (0.1177)	2.151^d (0.05717)	4.870^d (0.1362)	.9879	.9645	.9839	1,074	2,084	6
	Productive	1956	2.264^e (0.02680)	2.228^f (0.01617)	5.043^g (0.02483)	.9942	.9968	.9914	125.2	2,061	4
	Productive	1959	2.987^d (0.06099)	2.098^e (0.02655)	6.268^f (0.05138)	.9988	.9831	.9983	248.5	2,063	4
	Oil	1959	0.9403^d (0.04126)	3.885^d (0.09910)	3.653^e (0.06820)	.9995	.9800	.9990	144.2	881	2
	Gas	1959	5.224^g (0.01901)	1.510^g (0.003826)	7.888^h (0.009178)	1.0000	.9997	1.0000	40.40	1,182	4
Arkansas	Dry	1955	1.035^d (0.02897)	2.575^e (0.03517)	2.666^e (0.03932)	.9951	.9953	.9926	283.7	364	4
	Dry	1956	2.260^c (0.1413)	1.728^d (0.06726)	3.904^d (0.09668)	.9938	.9842	.9907	560.6	332	4
	Dry	1959	4.195^f (0.03565)	1.249^f (0.007368)	5.239^g (0.01427)	.9998	.9994	.9998	76.30	334	4
	Productive	1955	1.776^d (0.06041)	2.560^e (0.04438)	4.546^e (0.07857)	.9971	.9934	.9957	493.5	439	4
	Productive	1956	3.167^e (0.03340)	1.976^f (0.01237)	6.259^g (0.02838)	.9992	.9987	.9988	156.0	670	4
	Productive	1959	2.443^c (0.1833)	2.499^d (0.08721)	6.105^d (0.2542)	.9607	.9684	.9450	2,128	503	7
	Oil	1959	1.760^d (0.05948)	2.689^e (0.04070)	4.731^e (0.09137)	.9955	.9877	.9932	674.4	462	4
	Gas	1959	19.55^a (5.970)	0.9180^a (0.2085)	17.95^b (1.456)	.9908	.9501	.9862	7,968	41	4

Data area	Type of well	Year	K ($\times 10^4$ omitted)	α ($\times 10^{-4}$ omitted)	H (current dollars per foot)	I^2	J^2	$\bar{I}^2$	S (current dollars)	Wells	Degrees of freedom
Onshore California[l]	Dry	1956	8.510^c (0.7538)	1.118^c (0.05838)	9.510^d (0.3572)	.9817	.8244	.9765	3,712	568	7
	Dry	1959	7.734^b (0.8124)	1.100^c (0.06378)	8.505^d (0.4164)	.9746	.7305	.9674	5,879	510	7
	Productive	1956	66.79^a (20.96)	0.2650^a (0.07414)	17.70^d (0.6511)	.9895	0	.9865	4,789	1,722	7
	Productive	1959	14.44^d (0.3036)	0.9355^e (0.01250)	13.51^f (0.1061)	.9974	.9911	.9965	785.0	931	6
	Oil	1959	15.90^d (0.3766)	0.8842^e (0.01357)	14.06^f (0.1201)	.9974	.9903	.9965	848.0	857	6
	Gas	1959	22.93^a (6.317)	0.5658^a (0.1286)	12.97^c (0.6543)	.9939	.9493	.9914	3,681	74	5
Offshore California[l]	Dry[o]	1956	∞	0	29.95^a (10.95)	0	0	0	240,800	5	3
	Dry	1959	Insufficient data								
	Productive	1956	111.1 (245.3)	0.2641 (0.5286)	29.34^a (6.209)	.9819	.8782	.9639	20,486	9	2
	Productive	1959	5.298^b (1.034)	2.755^c (0.1885)	14.60^b (1.876)	.9800	.9807	.9701	20,270	27	4
	Oil	1959	5.298^b (1.034)	2.755^c (0.1885)	14.60^b (1.876)	.9800	.9807	.9701	20,270	27	4
	Gas	1959	No gas wells drilled								

Table B3 — Drilling Cost Regressions: Joint Association Survey Data as Published, 1955, 1956, and 1959 (continued)

Data area	Type of well	Year	K ($\times 10^4$ omitted)	α ($\times 10^{-4}$ omitted)	H (current dollars per foot)	$\bar{I}^2$	$\bar{J}^2$	$\bar{\bar{I}}^2$	S (current dollars)	Wells	Degrees of freedom
Total California	Dry	1955	9.873^c (0.6729)	1.098^d (0.04346)	10.84^d (0.3255)	.9847	.8524	.9803	4,245	644	7
	Dry	1956	10.48^c (0.9652)	0.9893^c (0.05664)	10.37^d (0.3752)	.9804	.8208	.9748	3,837	573	7
	Dry	1959	7.888^b (0.8237)	1.089^c (0.06320)	8.590^d (0.4148)	.9754	.7350	.9684	5,814	512	7
	Productive	1955	13.66^c (0.7474)	0.9121^d (0.03293)	12.46^e (0.2407)	.9909	.9460	.9883	2,208	1,811	7
	Productive	1956	69.02^a (22.08)	0.2585^a (0.07398)	17.84^d (0.6527)	.9898	0	.9869	4,790	1,731	7
	Productive	1959	15.36^d (0.3570)	0.9101^e (0.01363)	13.98^f (0.1187)	.9970	.9884	.9960	874.6	958	6
	Oil	1959	17.11^d (0.4402)	0.8535^h (0.001446)	14.60^i (0.01316)	.9970	.9879	.9960	925.8	884	6
	Gas	1959	22.93^a (6.318)	0.5658^a (0.1287)	12.97^d (0.6544)	.9939	.9493	.9914	3,681	74	5
Colorado	Dry	1955	0.1433^d (0.003745)	5.1231^g (0.02382)	0.7342^d (0.01590)	.9900	.9438	.9867	926.5	1,023	6
	Dry	1956	0.6377^b (0.07794)	3.521^d (0.1359)	2.245^c (0.1932)	.9506	.9484	.9342	4,747	864	6
	Dry	1959	2.424^c (0.2401)	2.178^d (0.09098)	5.279^c (0.3223)	.9702	.9417	.9603	6,206	560	6
	Productive	1955	1.148^c (0.09154)	2.877^d (0.1048)	3.302^d (0.1451)	.9801	.9375	.9702	1,263	486	4
	Productive	1956	2.527^a (0.5598)	2.194^b (0.2552)	5.545^b (0.5925)	.9394	.8810	.9152	4,477	358	5
	Productive	1959	2.204^d (0.1085)	2.702^d (0.0582)	5.953^d (0.1674)	.9988	.9906	.9984	1,799	248	5
	Oil	1959	3.974^c (0.3329)	1.647^c (0.08675)	6.546^d (0.2072)	.9965	.9716	.9947	1,059	160	4
	Gas	1959	6.457^c (0.4072)	1.810^d (0.06176)	11.69^d (0.3437)	.9994	.9904	.9992	2,759	88	5

Data area	Type of well	Year	K $(\times 10^4$ omitted)	α $(\times 10^{-4}$ omitted)	H (current dollars per foot)	I^2	J^2	$\bar{I}^2$	S (current dollars)	Wells	Degrees of freedom
Illinois	Dry[o]	1955	∞	0	$5.795^{d,s}$ (0.2373)	.9279	.6144	.9279	1,286	1,773	4
	Dry[o]	1956	∞	0	$5.641^{d,s}$ (0.1922)	.9587	.9536	.9587	914.2	2,066	3
	Dry	1959	1.707^f (0.01138)	2.155^g (0.01068)	3.678^h (0.006414)	1.0000	1.0000	.9999	9,092	1,062	2
	Productive[o]	1955	∞	0	$10.75^{a,q}$ (3.293)	0	.07695	0	17,100	2,116	3
	Productive	1956	1.612^d (0.03483)	3.200^e (0.04416)	5.160^f (0.04082)	.9999	.9888	.9998	69.85	1,737	2
	Productive	1959	3.420^d (0.1434)	2.055^d (0.06431)	7.028^e (0.07732)	.9998	.9946	.9996	174.2	1,018	3
	Oil	1959	3.415^d (0.1419)	2.057^d (0.06380)	7.025^e (0.07670)	.9998	.9947	.9996	173.4	1,009	3
	Gas	1959	Insufficient data[m]								
Indiana	Dry	1955	10.19^b (1.252)	0.5334^b (0.06166)	5.438^f (0.04239)	.9975	.9963	.9949	56.81	477	2
	Dry	1956	1.845^a (0.4857)	2.918^b (0.5408)	5.383^c (0.4363)	.9974	.5065	.9948	745.1	439	2
	Dry	1959	1.880^d (0.07991)	1.565^d (0.05593)	2.941^f (0.02213)	.9998	.9984	.9995	40.23	566	2
	Productive	1955	1.246^e (0.01296)	4.049^f (0.02636)	5.046^g (0.02015)	.9999	.9996	.9996	29.99	227	1
	Productive	1956	2.193^d (0.05910)	3.015^d (0.05644)	6.612^f (0.05596)	.9994	.9973	.9981	76.20	295	1
	Productive[o]	1959	∞	0	$10.73^{e,s}$ (0.1662)	.9964	.9956	.9964	471.7	306	2
	Oil[o]	1959	∞	0	$10.72^{e,s}$ (0.1558)	.9966	.9961	.9966	456.4	295	2
	Gas	1959	Insufficient data[m]								

Table B3 — Drilling Cost Regressions: Joint Association Survey Data as Published, 1955, 1956, and 1959 (continued)

Data area	Type of well	Year	K ($\times 10^4$ omitted)	α ($\times 10^{-4}$ omitted)	H (current dollars per foot)	I^2	J^2	$\bar{I}^2$	S (current dollars)	Wells	Degrees of freedom
Kansas	Dry	1955	1.208[e] (0.02537)	2.622[e] (0.03194)	3.167[f] (0.02838)	.9975	.9822	.9959	111.9	2,149	3
	Dry	1956	0.6375[d] (0.01947)	3.819[e] (0.05342)	2.435 (0.04089)	.9853	.9594	.9755	228.2	2,242	3
	Dry	1959	0.8501[f] (0.07113)	3.200[d] (0.1358)	2.720[d] (0.1147)	.9992	.8739	.9987	693.4	1,937	4
	Productive	1955	3.834[c] (0.2464)	1.718[d] (0.07991)	6.588[e] (0.1191)	.9894	.9220	.9822	296.9	2,796	3
	Productive	1956	2.284 (1.273)	2.503[a] (0.8513)	5.716[a] (1.266)	.9840	0	.9760	4,718	2,621	4
	Productive	1959	3.346[f] (0.02024)	1.949[g] (0.007774)	6.519[g] (0.01371)	.9997	.9985	.9995	45.24	1,943	3
	Oil	1959	3.364[f] (0.02206)	1.939[g] (0.008514)	6.521[g] (0.01445)	.9998	.9987	.9997	46.38	1,760	3
	Gas	1959	3.763[c] (0.2912)	1.818[d] (0.09006)	6.841[d] (0.1923)	.9980	.9775	.9960	492.4	183	2
Kentucky	Dry	1955	2.575[a] (0.5472)	1.887[b] (0.3107)	4.860[c] (0.2493)	.9944	.9186	.9908	515.2	774	3
	Dry	1956	1.156[a] (0.3375)	3.645[b] (0.4404)	4.214[b] (0.7718)	.9324	.8564	.8985	3,559	971	4
	Dry	1959	2.707[d] (0.09521)	1.842[d] (0.03907)	4.986[e] (0.08001)	.9901	.9958	.9862	296.1	893	5
	Productive[o]	1955	∞	0	8.558[d,s] (0.2443)	.9911	.9868	.9911	838.2	816	3
	Productive[o]	1956	∞	0	9.117[c,s] (0.5227)	.9410	.7581	.9410	1,874	960	4
	Productive	1959	31.55[a] (6.747)	0.2940[d] (0.06014)	9.277[e] (0.09573)	.9980	.9814	.9967	130.3	2,435	3
	Oil[o]	1959	∞	0	9.476[d,r] (0.4468)	.9648	.8910	.9648	630.0	2,146	2
	Gas	1959	10.52[d] (0.2406)	0.7985[e] (0.01571)	8.398[g] (0.02783)	.9998	.9994	.9997	63.98	289	3

Data area	Type of well	Year	K ($\times 10^4$ omitted)	α ($\times 10^{-4}$ omitted)	H (current dollars per foot)	I^2	J^2	$\bar{I}^2$	S (current dollars)	Wells	Degrees of freedom
North Louisiana[j]	Dry	1959	1.184[g] (0.002408)	2.474[i] (0.002112)	2.929[h] (0.003527)	.9999	1.0000	.9999	39.75	601	5
	Productive	1959	1.919[d] (0.04025)	2.467[f] (0.02057)	4.734[e] (0.06061)	.9962	.9956	.9947	497.8	916	5
	Oil	1959	3.354[d] (0.08257)	1.705[e] (0.02307)	5.719[e] (0.06640)	.9973	.9948	.9962	421.0	752	5
	Gas	1959	1.100[b] (0.1248)	3.148[d] (0.1130)	3.465[c] (0.2695)	.9901	.9777	.9861	3,211	164	5
South Louisiana[j]	Dry	1959	1.091[e] (0.01736)	2.444[g] (0.01086)	2.666[e] (0.03069)	.9994	.9941	.9992	1,294	706	7
	Productive	1959	3.504[e] (0.04240)	1.842[g] (0.007991)	6.457[f] (0.05043)	.9992	.9929	.9990	1,306	1,137	7
	Oil	1959	4.578[e] (0.06825)	1.638[f] (0.009844)	7.500[f] (0.06734)	.9988	.9928	.9985	1,439	818	7
	Gas	1959	2.580[d] (0.05644)	2.073[f] (0.01431)	5.347[e] (0.08036)	.9982	.9936	.9974	2,398	319	5
Onshore Louisiana[n]	Dry	1955	13.12[a] (2.982)	0.7647[b] (0.1124)	10.03[c] (0.8290)	.8601	0	.8202	10,410	1,080	7
	Dry	1956	1.998[e] (0.2863)	2.125[g] (0.009649)	4.245[f] (0.04183)	.9971	.9929	.9962	1,269	1,250	7
	Dry	1959	1.149[g] (0.004404)	2.410[h] (0.002633)	2.769[g] (0.007624)	.9997	.9996	.9997	253	1,307	7
	Productive	1955	4.101[e] (0.06181)	1.692[f] (0.01032)	6.938[f] (0.06296)	.9970	.9859	.9961	1,021	2,546	7
	Productive	1956	2.470[g] (0.006757)	2.176[i] (0.001899)	5.375[h] (0.01008)	.9998	.9997	.9997	256.4	2,196	7
	Productive	1959	3.504[g] (0.009495)	1.843[i] (0.001801)	6.459[h] (0.01127)	.9998	.9996	.9998	23.81	2,053	7
	Oil	1959	4.163[f] (0.02239)	1.699[g] (0.003596)	7.070[g] (0.02329)	.9996	.9989	.9995	412.6	1,570	7
	Gas	1959	3.011[e] (0.03664)	1.976[g] (0.007983)	5.950[f] (0.04862)	.9989	.9972	.9985	1,360	483	7

146

Table B3 – Drilling Cost Regressions: Joint Association Survey Data as Published, 1955, 1956, and 1959 (continued)

Data area	Type of well	Year	K (× 10^4 omitted)	α (× 10^-4 omitted)	H (current dollars per foot)	I^2	J^2	$\bar{I}^2$	S (Current dollars)	Wells	Degrees of freedom
Offshore Louisiana	Dry	1955	16.20[b] (2.381)	1.212[c] (0.09483)	19.64[c] (1.366)	.9881	.9099	.9833	16,480	96	5
	Dry	1956	7.905[d] (0.3162)	1.742[e] (0.02614)	13.77[d] (0.3465)	.9990	.9896	.9987	8,266	139	7
	Dry	1959	9.916[a] (2.165)	1.372[b] (0.1384)	13.61[b] (1.612)	.9808	.7148	.9754	27,100	118	7
	Productive	1955	1.895 (1.340)	2.988[b] (0.5723)	5.664 (2.934)	.9289	.5849	.9004	109,600	292	5
	Productive	1956	4.809[c] (0.2602)	2.184[e] (0.03986)	10.50[d] (0.3806)	.9892	.9722	.9837	9,890	287	4
	Productive	1959	6.730[c] (0.6243)	1.814[d] (0.06596)	12.21[c] (0.6960)	.9879	.8760	.9798	13,291	323	3
	Oil	1959	7.688[b] (1.009)	1.656[c] (0.09647)	12.73[c] (0.9400)	.9936	.8400	.9893	13,840	249	3
	Gas	1959	12.76[b] (2.496)	1.464[c] (0.1243)	18.67[b] (2.080)	.9801	.8547	.9668	29,910	74	3
Total Louisiana	Dry	1955	4.452[d] (0.1074)	1.588[e] (0.01612)	7.069[e] (0.09991)	.9948	.9724	.9933	1,914	1,176	7
	Dry	1956	2.070[f] (0.01195)	2.226[h] (0.003895)	4.607[g] (0.01865)	.9994	.9988	.9992	606.2	1,389	7
	Dry	1959	1.587[f] (0.009497)	2.257[h] (0.004109)	3.581[g] (0.01499)	.9994	.9988	.9992	446.6	1,425	7
	Productive	1955	6.121[e] (0.07649)	1.526[f] (0.008484)	9.340[f] (0.06570)	.9980	.9877	.9974	991.3	2,838	7
	Productive	1956	3.400[g] (0.01677)	2.034[h] (0.003438)	6.916[g] (0.02261)	.9993	.9986	.9991	532.3	2,483	7
	Productive	1959	5.801[f] (0.03251)	1.591[g] (0.003674)	9.231[g] (0.03073)	.9994	.9974	.9992	573.6	2,376	7
	Oil	1959	7.124[e] (0.07343)	1.419[g] (0.006726)	10.11[f] (0.05706)	.9972	.9941	.9964	891.8	1,819	7
	Gas	1959	5.182[e] (0.1048)	1.707[f] (0.01304)	8.847[e] (0.1120)	.9959	.9892	.9947	2,586	557	7

147

Data area	Type of well	Year	K $(\times 10^4$ omitted)	α $(\times 10^{-4}$ omitted)	H (current dollars per foot)	I^2	J^2	Υ^2	S (current dollars)	Wells	Degrees of freedom
Michigan	Dry	1955	0.09819^d (0.02500)	3.250^e (0.04274)	3.191^e (0.04005)	.9981	.9947	.9969	168.1	298	3
	Dry	1956	1.739^e (0.1606)	2.353^c (0.1300)	4.092^d (0.1606)	.9938	.9830	.9907	836.9	223	4
	Dry	1959	1.269^e (0.02276)	3.039^f (0.02908)	3.856^f (0.03302)	.9991	.9979	.9985	156.1	308	3
	Productive	1955	3.334^c (0.2679)	1.739^c (0.09566)	5.798^d (0.1507)	.9963	.9688	.9939	469.2	213	3
	Productive	1956	19.77^c (1.869)	0.4416^f (0.03816)	8.732^f (0.07382)	.9997	.9936	.9996	176.4	214	3
	Productive	1959	5.853^b (1.117)	1.621^b (0.2320)	9.488^d (0.4640)	.9964	.8191	.9939	1,110	295	3
	Oil	1959	6.494^a (1.379)	1.510^b (0.2443)	9.804^c (0.5080)	.9957	.8164	.9928	1,214	241	3
	Gas	1959	4.093^f (0.02206)	2.028^g (0.007866)	8.299^h (0.01278)	1.0000	1.0000	1.0000	23.58	54	2
Mississippi	Dry	1955	1.269^c (0.06658)	2.342^d (0.04435)	2.973^d (0.1010)	.9886	.9836	.9848	2,076	267	6
	Dry	1956	2.102^c (0.1076)	1.767^d (0.04092)	3.715^d (0.1060)	.9942	.9775	.9919	1,523	291	5
	Dry	1959	2.762^c (0.2709)	1.497^d (0.06794)	4.137^c (0.2202)	.9970	.8387	.9959	3,240	362	6
	Productive	1955	1.445^b (0.2183)	2.324^d (0.1102)	3.357^b (0.3512)	.9766	.9527	.9699	11,260	181	7
	Productive	1956	4.323^a (0.9869)	1.440^b (0.1706)	6.223^b (0.6960)	.9827	.8748	.9712	6,985	150	3
	Productive	1959	1.295^c (0.1102)	2.401^d (0.06335)	3.110^c (0.1834)	.9876	.9398	.9826	5,333	283	5
	Oil	1959	1.152^b (0.1223)	2.489^d (0.07963)	2.867^c (0.2139)	.9878	.9349	.9816	6,202	251	4
	Gas	1959	2.645^b (0.4532)	1.895^c (0.1198)	5.011^b (0.5461)	.9884	.9464	.9838	11,720	32	5

Data area	Type of well	Year	K ($\times 10^4$ omitted)	α ($\times 10^{-4}$ omitted)	H (current dollars per foot)	I^2	J^2	$\bar{I}^2$	S (current dollars)	Wells	Degrees of freedom
Montana	Dry	1955	3.887^b (0.4531)	1.825^c (0.1042)	7.093^c (0.4349)	.9817	.9464	.9756	4,614	225	6
	Dry	1956	5.393^b (0.6086)	1.452^c (0.09994)	7.828^d (0.3570)	.9888	.8779	.9843	2,624	253	5
	Dry	1959	2.228^a (0.7369)	2.304^b (0.3056)	5.132^a (1.036)	.9542	.8850	.9389	13,370	169	6
	Productive	1955	4.050^a (1.945)	1.765^a (0.4295)	7.148^a (1.707)	.9754	.8325	.9672	12,020	186	6
	Productive	1956	4.942 (4.594)	1.626^a (0.8095)	8.033^a (3.488)	.9754	.5428	.9655	18,770	241	7
	Productive	1959	36.11^a (9.242)	0.3462^d (0.07786)	12.50^d (0.4030)	.9986	.9654	.9980	1,895	179	5
	Oil	1959	16.91^b (1.702)	0.6414^c (0.05064)	10.84^d (0.2420)	.9997	.9864	.9996	1,242	168	5
	Gas	1959	12.72^b (2.519)	1.553^b (0.2264)	19.75^c (1.092)	.9982	.9953	.9947	3,005	11	1
Nebraska	Dry	1955	25.08^a (10.16)	0.1662^a (0.06420)	4.168^b (0.07929)	.9999	.9692	.9999	111.3	580	3
	Dry	1956	0.1593^a (0.05821)	4.557^b (0.5521)	0.7261^a (0.1776)	.9670	.8474	.9505	1,425	616	4
	Dry^o	1959	∞	0	$3.742^{d,s}$ (0.08904)	.5337	.9600	.5337	1,309	616	5
	Productive^o	1955	∞	0	$7.242^{d,s}$ (0.2446)	.5789	.9133	.5789	3,204	311	4
	Productive	1956	1.604 (1.564)	2.162 (1.155)	3.467^a (1.531)	.9830	.6341	.9716	3,797	302	3
	Productive^o	1959	∞	0	$7.672^{d,s}$ (0.2156)	.9795	.9868	.9795	710.5	295	2
	Oil	1959	$89,950^a$ (40,840)	0.0008530^a (0.0003871)	7.674^g (0.01652)	.9999	.9790	.9996	171.8	293	1
	Gas	1959	Insufficient data								

Table B3 — Drilling Cost Regressions: Joint Association Survey Data as Published, 1955, 1956, and 1959 (continued)

Data area	Type of well	Year	K ($\times 10^{-4}$ omitted)	α ($\times 10^{-4}$ omitted)	H (current dollars per foot)	I^2	J^2	$\bar{I}^2$	S (current dollars)	Wells	Degrees of freedom
Northwest New Mexico[j]	Dry	1959	13.44[b] (2.524)	0.8226[b] (0.1077)	11.05[c] (0.6833)	.9753	.9332	.9671	5,195	112	6
	Productive	1959	7.299[a] (1.841)	1.388[b] (0.2330)	10.13[c] (0.8816)	.9768	0	.9675	4,707	787	5
	Oil	1959	62.62 (69.57)	0.2003 (0.2108)	12.54[c] (0.7821)	.9928	.4827	.9900	2,686	410	5
	Gas	1959	3.089[d] (0.1050)	2.405[e] (0.04067)	7.430[e] (0.1283)	.9969	.9895	.9954	791.9	377	4
Southeast New Mexico	Dry	1959	11.95[d] (0.5645)	0.8795[d] (0.02684)	10.51[e] (0.1821)	.9959	.9757	.9946	1,590	302	6
	Productive	1959	11.10[d] (0.4534)	0.9378[d] (0.02598)	10.41[e] (0.1445)	.9951	.9771	.9934	1,232	879	6
	Oil	1959	10.92[c] (0.6094)	0.9455[d] (0.03567)	10.33[e] (0.1973)	.9925	.9640	.9900	1,680	829	6
	Gas	1959	11.69[a] (4.226)	0.9321[a] (0.2274)	10.90[b] (1.328)	.9809	.6406	.9732	11,050	50	5
Total New Mexico	Dry	1955	7.084[b] (0.9309)	1.203[c] (0.09134)	8.520[c] (0.4890)	.9851	.9391	.9801	4,392	240	6
	Dry	1956	9.303[a] (2.011)	0.9927[b] (0.1285)	9.235[c] (0.8295)	.9895	0	.9865	8,747	369	7
	Dry	1959	12.21[d] (0.5763)	0.8691[g] (0.02681)	10.61[e] (0.1806)	.9950	.9705	.9934	1,518	414	6
	Productive	1955	11.05[c] (0.9155)	0.9434[c] (0.05136)	10.42[d] (0.3117)	.9930	.9066	.9910	2,757	1,423	7
	Productive	1956	11.36[c] (0.7696)	0.9575[d] (0.04147)	10.87[d] (0.2799)	.9946	.9444	.9931	2,644	1,535	7
	Productive	1959	12.51[d] (0.5496)	0.8894[d] (0.02800)	11.13[e] (0.1478)	.9896	.9616	.9861	1,154	1,666	6
	Oil	1959	11.27[c] (0.6383)	0.9218[g] (0.03632)	10.39[e] (0.1905)	.9926	.9585	.9902	1,488	1,239	6
	Gas	1959	9.367[a] (2.249)	1.198[b] (0.1932)	11.22[c] (0.9247)	.9864	0	.9818	7,290	427	6

Data area	Type of well	Year	K ($\times 10^4$ omitted)	α ($\times 10^{-4}$ omitted)	H (current dollars per foot)	I^2	J^2	$\bar{I}^2$	S (current dollars)	Wells	Degrees of freedom
North Dakota	Dry	1955	1.259^c (0.1004)	2.532^d (0.07289)	3.188^c (0.1651)	.9952	.9915	.9932	3,105	85	5
	Dry	1956	0.4184^e (0.04492)	3.566^d (0.09508)	1.492^c (0.1213)	.9703	.9914	.9604	4,466	99	6
	Dry	1959	1.883^d (0.05667)	2.065^e (0.02763)	3.889^e (0.06596)	.9984	.9943	.9976	700.4	158	4
	Productive	1955	1.264^a (0.5090)	2.635^b (0.4163)	3.330^a (0.8200)	.9915	.8329	.9880	6,314	165	5
	Productive	1956	2.780^b (0.3782)	1.822^c (0.1182)	5.068^c (0.3647)	.9958	.9606	.9941	4,531	164	5
	Productive	1959	2.331^c (0.1421)	2.149^d (0.05411)	5.011^d (0.1813)	.9938	.9821	.9908	2,321	278	4
	Oil	1959	2.448^c (0.1259)	2.098^d (0.04610)	5.136^d (0.1532)	.9937	.9887	.9906	1,868	273	4
	Gas	1959	Insufficient data								
Oklahoma	Dry	1955	1.973^d (0.06536)	2.116^e (0.03073)	4.176^d (0.08039)	.9863	.9637	.9817	818.3	2,588	6
	Dry	1956	2.608^b (0.3534)	1.880^c (0.1174)	4.905^c (0.3742)	.9840	.8794	.9794	4,275	2,476	7
	Dry	1959	1.539^e (0.01667)	2.388^g (0.007412)	3.674^f (0.02894)	.9987	.9990	.9983	582.6	2,072	7
	Productive	1955	3.588^d (0.1215)	1.741^e (0.02981)	6.247^e (0.1078)	.9795	.9558	.9727	905.6	5,490	6
	Productive	1956	2.555^d (0.09208)	2.108^e (0.03232)	5.386^d (0.1148)	.9723	.9728	.9644	1,398	5,146	7
	Productive	1959	2.668^d (0.09730)	2.206^e (0.03217)	5.884^d (0.1331)	.9959	.9697	.9947	2,115	3,175	7
	Oil	1959	2.846^b (0.4918)	2.092^c (0.1708)	5.953^c (0.5549)	.9947	.7699	.9929	4,896	2,677	6
	Gas	1959	3.230^d (0.1522)	2.097^e (0.03687)	6.773^d (0.2059)	.9930	.9732	.9909	4,945	498	7

Data area	Type of well	Year	K ($\times 10^4$ omitted)	α ($\times 10^{-4}$ omitted)	H (current dollars per foot)	I^2	J^2	$\bar{I}^2$	S (current dollars)	Wells	Degrees of freedom
East Texas	Dry	1955	0.8673[c] (0.08568)	2.944[d] (0.09540)	2.554[c] (0.1730)	.9831	.9216	.9774	3,688	479	6
	Dry	1956	1.688[b] (0.3236)	2.146[c] (0.1744)	3.622[b] (0.4157)	.9706	.2862	.9622	7,277	564	7
	Dry	1959	0.9936[d] (0.02686)	2.663[f] (0.02662)	2.646[e] (0.04592)	.9984	.9964	.9978	788.6	429	6
	Productive	1955	2.110[c] (0.1056)	2.407[d] (0.04856)	5.079[d] (0.1540)	.9993	.9802	.9991	2,071	505	6
	Productive	1956	1.791[c] (0.1436)	2.453[d] (0.07901)	4.394[d] (0.2169)	.9911	.9343	.9881	3,770	567	6
	Productive	1959	1.336[d] (0.04651)	2.713[e] (0.03357)	3.625[d] (0.08265)	.9854	.9906	.9805	1,271	813	6
	Oil	1959	1.233[b] (0.1288)	2.796[d] (0.1056)	3.448[c] (0.2345)	.9681	.9704	.9574	3,349	715	6
	Gas	1959	1.403[d] (0.02831)	2.666[f] (0.01801)	3.741[e] (0.05076)	.9997	.9990	.9995	1,015	98	4
Gulf Coast Texas	Dry	1955	0.9307[d] (0.02642)	2.654[f] (0.02253)	2.470[d] (0.04990)	.9933	.9879	.9914	1,732	1,100	7
	Dry	1956	1.125[d] (0.03579)	2.481[e] (0.02671)	2.792[d] (0.05958)	.9864	.9853	.9825	1,622	1,098	7
	Dry	1959	0.5144[d] (0.01920)	3.251[f] (0.03038)	1.672[d] (0.04717)	.9924	.9943	.9903	2,118	902	7
	Productive	1955	1.982[d] (0.07299)	2.303[e] (0.03214)	4.563[d] (0.1068)	.9943	.9754	.9927	2,399	1,390	7
	Productive	1956	2.446[d] (0.1164)	2.070[e] (0.04032)	5.064[d] (0.1458)	.9917	.9480	.9893	3,098	1,498	7
	Productive	1959	1.360[e] (0.02467)	2.641[f] (0.01449)	3.591[e] (0.04614)	.9935	.9942	.9916	1,576	1,093	7
	Oil	1959	1.434[c] (0.1174)	2.609[d] (0.06946)	3.742[c] (0.2105)	.9780	.9622	.9718	5,527	681	7
	Gas	1959	1.654[e] (0.02736)	2.499[f] (0.01276)	4.135[e] (0.04788)	.9983	.9976	.9978	1,595	412	6

Data area	Type of well	Year	K $(\times 10^4$ omitted)	a $(\times 10^{-4}$ omitted)	H (current dollars per foot)	I^2	J^2	$\bar{I}^2$	S (current dollars)	Wells	Degrees of freedom
North Central Texas	Dry	1955	1.088[g] (0.005339)	2.723[g] (0.005647)	2.962[g] (0.008674)	.9984	.9994	.9977	81.13	3,034	5
	Dry	1956	1.382[h] (0.002002)	2.298[f] (0.01691)	3.176[f] (0.02363)	.9969	.9930	.9957	179.6	3,417	5
	Dry	1959	1.136[b] (0.1266)	2.555[c] (0.1375)	2.902[c] (0.1766)	.9948	.7203	.9930	1,668	2,553	6
	Productive	1955	2.756[e] (0.05278)	1.972[e] (0.02045)	5.436[f] (0.04992)	.9938	.9800	.9913	375.4	4,398	5
	Productive	1956	3.084[e] (0.04930)	1.839[f] (0.01697)	5.673[f] (0.04026)	.9976	.9950	.9968	302.1	4,184	6
	Productive	1959	5.356[c] (0.4018)	1.256[c] (0.06286)	6.729[d] (0.1865)	.9859	.7713	.9802	1,255	3,298	5
	Oil	1959	2.437[c] (0.2152)	2.141[c] (0.1137)	5.218[d] (0.1894)	.9874	.9052	.9811	768.8	3,093	4
	Gas	1959	9.326[e] (0.1004)	0.9014[f] (0.006370)	8.406[g] (0.03277)	.9997	.9995	.9996	264.6	205	4
Panhandle Texas	Dry	1955	2.337[c] (0.1247)	2.102[d] (0.04771)	4.912[d] (0.1535)	.9977	.9935	.9969	2,156	95	6
	Dry	1956	3.790[b] (0.5090)	1.690[c] (0.1154)	6.404[c] (0.4278)	.9931	.9619	.9903	3,766	153	5
	Dry	1959	15.36[a] (3.304)	0.5809[b] (0.1002)	8.922[d] (0.3892)	.9932	.9400	.9904	1,932	221	5
	Productive	1955	2.476[b] (0.2984)	2.432[c] (0.1232)	6.022[c] (0.4422)	.9901	.7601	.9861	4,548	960	5
	Productive	1956	5.452[b] (0.5978)	1.524[c] (0.1001)	8.311[d] (0.3827)	.9872	.7168	.9829	2,869	1,169	6
	Productive	1959	6.246[c] (0.3934)	1.354[d] (0.05090)	8.460[d] (0.2225)	.9923	.9633	.9892	1,639	1,039	5
	Oil	1959	10.36[b] (1.177)	0.9475[c] (0.07862)	9.814[d] (0.3151)	.9954	.9150	.9932	1,657	762	4
	Gas	1959	4.755[d] (0.1802)	1.586[e] (0.03023)	7.540[e] (0.1450)	.9948	.9935	.9927	1,334	277	5

Table B3 – Drilling Cost Regressions: Joint Association Survey Data as Published, 1955, 1956, and 1959 (continued)

Data area	Type of well	Year	K ($\times 10^4$ omitted)	α ($\times 10^{-4}$ omitted)	H (current dollars per foot)	I^2	J^2	$\bar{I}^2$	S (current dollars)	Wells	Degrees of freedom
Southwest Texas	Dry	1955	1.189^c (0.1068)	2.556^d (0.07656)	3.038^c (0.1864)	.9876	.9519	.9841	3,885	1,220	7
	Dry	1956	1.160^d (0.05518)	2.628^e (0.03618)	3.049^d (0.1051)	.9854	.9825	.9812	2,595	1,287	7
	Dry	1959	0.9311^d (0.02897)	2.752^f (0.02426)	2.562^d (0.05797)	.9954	.9944	.9941	1,437	1,131	7
	Productive	1955	1.279^e (0.01846)	2.879^f (0.01532)	3.704^f (0.03434)	.9948	.9953	.9928	384.8	1,860	5
	Productive	1956	1.954^b (0.2095)	2.335^d (0.1010)	4.562^c (0.3002)	.9902	.9256	.9874	4,296	1,777	7
	Productive	1959	3.469^c (0.2650)	1.928^d (0.06614)	6.688^d (0.2906)	.9897	.8907	.9868	4,010	1,274	7
	Oil	1959	1.086^b (0.1156)	3.160^d (0.1148)	3.430^c (0.2431)	.9630	.9455	.9482	2,473	917	5
	Gas	1959	4.850^c (0.3229)	1.656^d (0.05324)	8.032^d (0.2841)	.9977	.9608	.9970	4,548	357	7
West Texas	Dry	1955	5.209^d (0.1581)	1.272^e (0.02096)	6.626^e (0.09467)	.9924	.9611	.9899	1,070	954	6
	Dry	1956	6.272^d (0.2228)	1.211^e (0.02395)	7.594^e (0.1238)	.9932	.9711	.9912	1,495	939	7
	Dry	1959	1.737^d (0.04086)	2.243^f (0.01464)	3.897^e (0.06712)	.9912	.9960	.9887	1,847	1,216	7
	Productive	1955	8.029^b (1.022)	1.098^c (0.08322)	8.814^c (0.4690)	.9937	.7196	.9919	5,002	3,949	7
	Productive	1956	8.088^e (0.1012)	1.1324^c (0.08430)	9.159^g (0.04824)	.9950	.9699	.9933	483.4	4,768	6
	Productive	1959	5.376^d (0.1301)	1.488^e (0.01897)	8.000^e (0.09462)	.9981	.9690	.9975	1,110	4,163	7
	Oil	1959	6.202^f (0.03360)	1.356^g (0.004188)	8.410^g (0.02032)	.9986	.9965	.9982	203.8	4,035	6
	Gas	1959	6.684^d (0.2970)	1.455^d (0.02993)	9.723^d (0.2354)	.9981	.9888	.9976	3,656	128	7

Data area	Type of well	Year	K ($\times 10^4$ omitted)	α ($\times 10^{-4}$ omitted)	H (current dollars per foot)	$\bar{r}^2$	J^2	$\bar{I}^2$	(current dollars)	Wells	Degrees of freedom
Offshore Texas	Dry	1955	28.07 (37.93)	1.094 (0.8474)	30.72 (17.89)	.9363	.7346	.8725	134,300	9	2
	Dry	1956	56.19 (51.00)	0.6761 (0.4175)	37.99[a] (11.40)	.8638	0	.7956	126,600	26	4
	Dry	1959	Insufficient data								
	Productive	1955	1.260[c] (0.08162)	3.751[e] (0.0517)	4.725[c] (0.2417)	.9995	.9992	.9986	10,620	17	1
	Productive	1956	0.2391 (1.443)	4.952 (4.983)	1.184 (5.957)	.7213	.5763	.1640	299,110	13	1
	Productive	1959	Insufficient data								
	Oil	1959	Insufficient data								
	Gas	1959	Insufficient data								
Total Texas	Dry	1955	1.536[d] (0.04160)	2.273[f] (0.02252)	3.492[e] (0.06124)	.9870	.9736	.9833	1,069	6,891	7
	Dry	1956	1.590[e] (0.02032)	2.287[g] (0.01049)	3.636[f] (0.03046)	.9947	.9907	.9932	553.9	7,484	7
	Dry	1959	1.240[g] (0.005651)	2.484[h] (0.003243)	3.081[g] (0.01022)	.9977	.9995	.9971	259.4	6,453	7
	Productive	1955	4.211[d] (0.2024)	1.606[d] (0.03846)	6.766[d] (0.1688)	.9882	.8417	.9849	2,000	13,079	7
	Productive	1956	4.388[d] (0.1598)	1.578[e] (0.02931)	6.923[d] (0.1281)	.9935	.9039	.9917	1,490	13,976	7
	Productive	1959	3.356[d] (0.06950)	1.880[f] (0.01746)	6.308[e] (0.07401)	.9950	.9652	.9936	958.4	11,682	7
	Oil	1959	3.826[c] (0.3208)	1.740[d] (0.07292)	6.657[d] (0.2885)	.9956	.8368	.9944	3,087	10,204	7
	Gas	1959	3.190[d] (0.07698)	1.955[f] (0.01885)	6.237[e] (0.09204)	.9944	.9832	.9928	1,816	1,478	7

Table B3 – Drilling Cost Regressions: Joint Association Survey Data as Published, 1955, 1956, and 1959 (continued)

Data area	Type of well	Year	K (× 10^4 omitted)	α (× 10^-4 omitted)	H (current dollars per foot)	I^2	J^2	ι^2	S (current dollars)	Wells	Degrees of freedom
Utah	Dry	1955	12.64[b] (1.442)	1.145[c] (0.08027)	14.47[d] (0.6630)	.9938	.9706	.9918	5,518	68	6
	Dry	1956	31.88[a] (6.422)	0.7052[b] (0.1127)	22.48[d] (0.9740)	.9936	0	.9911	4,780	82	5
	Dry	1959	9.798[a] (1.934)	1.252[b] (0.1588)	12.26[c] (0.8979)	.9906	.9325	.9875	7,870	119	6
	Productive	1955	7.656[b] (0.8840)	1.867[c] (0.1041)	14.29[c] (0.8730)	.9953	.9868	.9935	9,538	24	5
	Productive[o]	1956	∞	0	29.67[d,s] (0.8175)	.9517	.9347	.9341	11,290	59	4
	Productive	1959	4.401[a] (1.170)	2.103[b] (0.2718)	9.257[e] (0.1332)	.9732	.7452	.9624	19,540	199	5
	Oil	1959	4.228[a] (1.404)	2.164[b] (0.3427)	9.150[b] (1.677)	.9678	.6436	.9517	24,240	183	4
	Gas	1959	7.173[c] (0.5412)	1.473[d] (0.06349)	10.56[d] (0.3538)	.9987	.9982	.9978	2,462	16	3
Wyoming	Dry	1955	4.038[c] (0.3257)	1.629[d] (0.06670)	6.577[d] (0.2717)	.9708	.9493	.9624	3,543	431	7
	Dry	1956	3.537[c] (0.3130)	1.948[d] (0.06561)	6.891[c] (0.3884)	.9815	.9609	.9762	6,581	430	7
	Dry	1959	2.357[c] (0.1619)	2.241[d] (0.06424)	5.281[d] (0.2160)	.9864	.9677	.9818	3,223	501	6
	Productive	1955	7.218[b] (0.7398)	1.429[c] (0.08364)	10.31[d] (0.4675)	.9824	.8609	.9765	4,527	484	6
	Productive	1956	5.791[d] (0.2121)	1.663[e] (0.03240)	9.630[d] (0.1718)	.9871	.9855	.9828	1,773	500	6
	Productive	1959	5.732[c] (0.3558)	1.620[d] (0.05166)	9.287[d] (0.2934)	.9683	.9789	.9592	4,128	448	7
	Oil	1959	203.6 (917.1)	0.07174 (0.3157)	14.61[b] (1.556)	.9410	.6905	.9174	5,060	371	5
	Gas	1959	10.66[c] (0.6915)	1.307[d] (0.04509)	13.93[d] (0.4407)	.9949	.9883	.9934	7,149	77	7

Table B3 — Drilling Cost Regressions: Joint Association Survey Data as Published, 1955, 1956, and 1959 (continued)

Footnotes

[a] Coefficient greater than 2 times asymptotic standard error.
[b] Coefficient greater than 5 times asymptotic standard error.
[c] Coefficient greater than 10 times asymptotic standard error.
[d] Coefficient greater than 20 times asymptotic standard error.
[e] Coefficient greater than 50 times asymptotic standard error.
[f] Coefficient greater than 100 times asymptotic standard error.
[g] Coefficient greater than 200 times asymptotic standard error.
[h] Coefficient greater than 500 times asymptotic standard error.
[i] Coefficient greater than 1,000 times asymptotic standard error.
[j] Not reported separately before 1959.
[k] Includes New York, Ohio, Pennsylvania, and West Virginia.
[l] Not reported separately before 1956.
[m] One depth class reported; none in JAS sample.
[n] Not reported as such by JAS for 1959; constructed for reasons of comparability by addition of corresponding figures for North and South Louisiana.
[o] Linear regression performed; see Appendix A. In these cases $\alpha = 0$, $K = \infty$, and the coefficient of the regression is given as H. I^2 and I^2 are
    identical by definition (since the number of degrees of freedom lost in both numerator and denominator of the relevant fraction is 1).
[p] Coefficient in straight line regression significant at 5% level.
[q] Coefficient in straight line regression significant at 2% level.
[r] Coefficient in straight line regression significant at 1% level.
[s] Coefficient in straight line regression significant at 1/10% level.

N. B. These significance levels should be interpreted with care. The slope of a straight line forced through the origin will be significantly
       different from zero roughly so long as the point of means of the observations lies away from the axes, regardless of the goodness of
       fit of the line. (See Illinois, productive wells, 1955, for example.)

Table B4 – Comparisons of Joint Association Survey Sample and Data as Published, 1959*

Data area and type of well	i_1^2/i_0^2	S_1/S_0	K_1/K_0	α_1/α_0	$a_1 \cdot \alpha_0$ ($\times 10^{-4}$ omitted)	H_1/H_0	MC_1/MC_0 5,000 ft.	MC_1/MC_0 10,000 ft.	MC_1/MC_0 15,000 ft.	Y_1/Y_0 5,000 ft.	Y_1/Y_0 10,000 ft.	Y_1/Y_0 15,000 ft.
Alabama												
Dry	1.010	0.346	0.907	1.058	0.047	0.960	0.982	1.006	1.030	0.891	0.986	1.001
Productive	Insufficient data											
Oil	Insufficient data											
Gas	Insufficient data											
Alaska												
Dry	2.669	12.90	Indeterminate	Indeterminate	0	1.071	1.071	1.071	1.071	1.071	1.071	1.071
Productive	0.995	0.841	Indeterminate	Indeterminate	0	1.090	1.090	1.090	1.090	1.090	1.090	1.090
Oil	Insufficient data											
Gas	Insufficient data											
Appalachian												
Dry	1.021	0.056	1.895	0.786	− 0.636	1.489	1.083	0.788	0.574	1.262	0.953	0.716
Productive	0.998	0.693	0.788	1.161	0.291	0.915	1.058	1.224	1.416	0.995	1.106	1.251
Oil	1.001	0.588	2.990	0.560	− 3.046	1.676	0.366	0.080	0.017	0.576	0.140	0.030
Gas	1.002	0.093	1.248	0.885	− 0.196	1.105	1.001	0.908	0.823	1.045	0.976	0.903
Arkansas												
Dry	1.034	0.0148	1.432	0.802	− 0.308	1.149	0.985	0.844	0.724	1.023	0.952	0.846
Productive	0.989	0.292	0.616	1.189	0.397	0.732	0.893	1.089	1.328	0.824	0.959	1.055
Oil	1.007	0.234	1.048	0.949	− 0.145	0.994	0.925	0.860	0.800	0.951	0.898	0.840
Gas	1.074	0.416	1.274	0.875	− 0.131	1.115	1.045	0.978	0.916	1.079	1.032	0.984

* Throughout this table, the subscript 1 denotes estimates from the published data and the subscript 0 denotes estimates from the sample.

Table B4 — Comparisons of Joint Association Survey Sample and Data as Published, 1959* (continued)

Data area and type of well	i_1^2/i_1^2	S_1/S_0	K_1/K_0	α_1/α_0	$\alpha_1-\alpha_0$ ($\times10^{-4}$ omitted)	H_1/H_0	MC_1/MC_0 5,000 ft.	MC_1/MC_0 10,000 ft.	MC_1/MC_0 15,000 ft.	Y_1/Y_0 5,000 ft.	Y_1/Y_0 10,000 ft.	Y_1/Y_0 15,000 ft.
Onshore California												
Dry	1.010	0.620	0.780	1.140	0.136	0.889	0.952	1.018	1.089	0.922	0.964	1.010
Productive	1.001	0.846	0.998	1.003	0.003	1.001	1.002	1.004	1.006	1.000	1.002	1.004
Oil	1.000	0.866	1.038	0.979	-0.019	1.016	1.006	0.997	0.988	1.012	1.005	0.998
Gas	1.000	0.742	1.011	0.982	-0.010	0.993	0.987	0.982	0.977	0.989	0.985	0.983
Offshore California												
Dry	Insufficient data											
Productive	1.000	0.997	0.999	1.000	0.001	0.999	1.000	1.000	1.001	1.001	1.001	1.001
Oil	1.000	0.997	0.999	1.000	0.001	0.999	1.000	1.000	1.001	1.001	1.001	1.001
Gas	No gas wells drilled											
Total California												
Dry	1.008	0.634	0.758	1.158	0.149	0.878	0.946	1.019	1.098	0.913	0.959	1.009
Productive	1.001	0.823	0.973	1.016	0.014	0.989	0.996	1.003	1.010	0.996	0.998	1.002
Oil	1.001	0.830	1.015	0.989	-0.009	1.004	0.999	0.995	0.990	1.002	1.000	0.996
Gas	1.000	0.742	1.149	0.982	-0.010	0.993	0.987	0.982	0.977	0.989	0.985	0.983
Colorado												
Dry	0.976	1.992	4.344	0.552	-1.769	2.337	0.989	0.409	0.169	1.384	0.670	0.294
Productive	1.010	0.270	1.400	0.873	-0.392	1.223	1.005	0.826	0.679	1.083	0.925	0.772
Oil	1.045	0.104	1.181	0.905	-0.172	1.069	0.981	0.899	0.826	1.018	0.958	0.893
Gas	1.006	0.462	1.149	0.953	-0.090	1.095	1.047	1.001	0.956	1.067	1.033	0.995

Table B4 – Comparisons of Joint Association Survey Sample and Data as Published, 1959* (continued)

Data area and type of well	I_1^2/I_0^2	S_1/S_0	K_1/K_0	α_1/α_0	$\alpha_1 \cdot \alpha_0$ ($\times 10^{-4}$ omitted)	H_1/H_0	MC_1/MC_0 5,000 ft.	MC_1/MC_0 10,000 ft.	MC_1/MC_0 15,000 ft.	Y_1/Y_0 5,000 ft.	Y_1/Y_0 10,000 ft.	Y_1/Y_0 15,000 ft.
Illinois												
Dry	1.005	0.034	0.666	1.398	0.614	0.930	1.265	1.720	2.314	1.111	1.384	1.782
Productive	1.004	0.336	1.025	0.984	−0.034	1.008	0.991	0.974	0.958	0.998	0.986	0.972
Oil	1.004	0.335	1.023	0.985	−0.032	1.007	0.991	0.975	0.960	0.997	0.987	0.973
Gas	No data in sample											
Indiana												
Dry	1.099	0.046	0	∞	1.565	0.836	1.828	3.998	8.738	1.268	2.021	3.369
Productive	0.998	1.232	Indeterminate	Indeterminate	0	1.032	1.032	1.032	1.032	1.032	1.032	1.032
Oil	0.998	1.192	Indeterminate	Indeterminate	0	1.031	1.031	1.031	1.031	1.031	1.031	1.031
Gas	No data in sample											
Kansas												
Dry	1.011	2.250	1.971	0.759	−1.017	1.496	0.900	0.541	0.326	1.077	0.694	0.426
Productive	1.038	0.068	6.528	0.405	−2.860	2.646	0.633	0.152	0.036	1.070	0.323	0.085
Oil	1.011	0.107	1.500	0.803	−0.475	1.205	0.950	0.749	0.591	1.048	0.877	0.716
Gas	1.018	0.242	1.661	0.748	−0.612	1.243	0.915	0.674	0.500	1.040	0.827	0.636
Kentucky												
Dry	0.996	0.594	2.140	0.625	−2.330	1.337	0.417	0.130	0.041	0.962	0.628	0.378
Productive	1.000	0.590	3.898	0.299	−0.688	1.166	0.826	0.586	0.415	0.973	0.797	0.642
Oil	0.983	0.651	Indeterminate	Indeterminate	0	1.085	1.085	1.085	1.085	1.085	1.085	1.085
Gas	1.002	0.142	1.741	0.638	−0.452	1.111	0.895	0.707	0.566	0.990	0.851	0.728

Table B4 – Comparisons of Joint Association Survey Sample and Data as Published, 1959* (continued)

Data area and type of well	$\bar{i}_1^2/\bar{i}_1^2$	S_1/S_0	K_1/K_0	α_1/α_0	$\alpha_1-\alpha_0$ ($\times 10^{-4}$ omitted)	H_1/H_0	MC_1/MC_0 5,000 ft.	MC_1/MC_0 10,000 ft.	MC_1/MC_0 15,000 ft.	Y_1/Y_0 5,000 ft.	Y_1/Y_0 10,000 ft.	Y_1/Y_0 15,000 ft.
North Louisiana												
Dry	1.002	0.029	1.333	0.900	− 0.276	1.199	1.044	0.910	0.792	1.103	0.990	0.874
Productive	1.015	0.144	3.697	0.637	− 1.406	2.355	1.166	0.577	0.286	1.516	0.847	0.439
Oil	1.004	0.178	1.203	0.906	− 0.176	1.090	0.999	0.914	0.837	1.037	0.974	0.906
Gas	0.995	0.242	4.315	0.674	− 1.519	2.910	1.361	0.637	0.298	1.775	0.914	0.437
South Louisiana												
Dry	1.001	0.341	1.036	0.991	− 0.023	1.026	1.014	1.003	0.991	1.019	1.010	1.000
Productive	1.000	0.571	1.129	0.956	− 0.085	1.079	1.034	0.991	0.950	1.054	1.022	0.987
Oil	1.003	0.466	1.071	0.980	− 0.033	1.050	1.033	1.015	0.999	1.040	1.028	1.015
Gas	1.004	0.510	1.018	0.992	− 0.016	1.010	1.002	0.994	0.986	1.005	1.000	0.993
Onshore Louisiana												
Dry	1.001	0.130	1.059	0.984	− 0.038	1.043	1.023	1.004	0.985	1.031	1.016	0.999
Productive	1.000	0.020	1.085	0.968	− 0.060	1.052	1.021	0.991	0.961	1.033	1.011	0.989
Oil	1.003	0.201	1.024	0.998	− 0.004	1.022	1.020	1.018	1.016	1.021	1.020	1.018
Gas	1.002	0.437	1.021	0.991	− 0.018	1.012	1.003	0.994	0.985	1.008	1.000	0.993
Offshore Louisiana												
Dry	0.989	1.376	1.261	0.903	− 0.148	1.138	1.057	0.982	0.911	1.093	1.040	0.982
Productive	1.002	1.004	1.037	0.994	− 0.010	1.031	1.029	1.021	1.016	1.028	1.025	1.020
Oil	1.001	0.986	0.990	1.005	0.009	0.995	0.996	1.004	1.008	1.000	1.001	1.004
Gas	0.968	0.609	1.262	0.917	− 0.132	1.158	1.084	1.015	0.950	1.114	1.067	1.015

Table B4 – Comparisons of Joint Association Survey Sample and Data as Published, 1959* (continued)

Data area and type of well	I_1^2/I_0^2	S_1/S_0	K_1/K_0	α_1/α_0	$\alpha_1-\alpha_0$ ($\times 10^4$ omitted)	H_1/H_0	MC_1/MC_0 5,000 ft.	MC_1/MC_0 10,000 ft.	MC_1/MC_0 15,000 ft.	Y_1/Y_0 5,000 ft.	Y_1/Y_0 10,000 ft.	Y_1/Y_0 15,000 ft.
Total Louisiana												
Dry	1.005	0.074	0.790	1.078	0.163	0.852	0.924	1.002	1.088	0.860	0.914	0.937
Productive	1.000	0.466	0.883	1.017	0.026	0.898	0.910	0.921	0.933	0.904	0.912	0.922
Oil	1.004	0.415	0.775	1.083	0.109	0.840	0.887	0.936	0.989	0.866	0.898	0.935
Gas	1.004	0.510	1.099	0.957	-0.077	1.052	1.013	0.974	0.937	1.029	1.001	0.970
Michigan												
Dry	1.028	0.067	0.069	8.269	2.672	0.572	2.173	8.270	31.46	1.225	3.089	8.870
Productive	1.048	0.119	1.438	0.721	-0.527	1.037	0.758	0.554	0.405	0.865	0.689	0.530
Oil	1.055	0.124	1.814	0.620	-0.408	1.124	0.916	0.748	0.609	0.987	0.614	0.416
Gas	Insufficient data[a]											
Mississippi												
Dry	1.048	0.251	1.739	0.788	-0.405	1.370	1.119	0.914	0.746	1.219	1.058	0.899
Productive	1.025	0.436	1.763	0.852	-0.418	1.502	1.219	0.989	0.802	1.323	1.122	0.930
Oil	1.021	0.420	2.140	0.811	-0.580	1.736	1.299	0.972	0.723	1.453	1.152	0.884
Gas	1.049	0.371	0.717	1.127	0.214	0.808	0.900	1.002	1.115	0.863	0.928	1.012
Montana												
Dry	1.021	0.699	1.937	0.804	-0.560	1.558	1.178	0.890	0.673	1.316	1.056	0.821
Productive	1.000	0.624	1.264	0.826	-0.073	1.045	1.008	0.972	0.938	1.024	1.006	0.995
Oil	1.000	0.798	0.837	1.154	0.086	0.966	1.008	1.052	1.088	0.985	1.012	1.036
Gas	Insufficient data[a]											

Table B4 — Comparisons of Joint Association Survey Sample and Data as Published, 1959* (continued)

Data area and type of well	i_1^2/i_0^2	S_1/S_0	K_1/K_0	α_1/α_0	$\alpha_1-\alpha_0$ (× 10^{-4} omitted)	H_1/H_0	MC_1/MC_0 5,000 ft.	MC_1/MC_0 10,000 ft.	MC_1/MC_0 15,000 ft.	Y_1/Y_0 5,000 ft.	Y_1/Y_0 10,000 ft.	Y_1/Y_0 15,000 ft.
Nebraska												
Dry	0.646	0.369	Indeterminate	Indeterminate	0	1.131	1.131	1.131	1.131	1.131	1.131	1.131
Productive	Insufficient data[a]											
Oil	Insufficient data[a]											
Gas	Insufficient data											
Northwest New Mexico												
Dry	1.130	0.306	0.610	1.384	0.228	0.844	0.946	1.060	1.189	0.896	0.959	1.032
Productive	1.026	0.651	1.172	0.911	−0.136	1.067	0.977	0.931	0.870	1.030	0.981	0.931
Oil	1.008	0.443	1.105	0.919	−0.018	1.015	1.006	0.997	0.988	1.008	1.008	1.001
Gas	0.997	0.819	1.068	0.958	−0.105	1.023	0.971	0.922	0.859	0.992	0.947	0.909
Southeast New Mexico												
Dry	1.007	0.408	1.071	0.952	−0.044	1.019	0.997	0.975	0.953	1.008	0.994	0.979
Productive	1.015	0.442	1.086	0.956	−0.043	1.039	1.017	0.996	0.975	1.027	1.015	0.999
Oil	1.020	0.495	1.085	0.963	−0.079	1.045	1.004	0.965	0.928	1.037	1.023	1.010
Gas	1.002	1.177	2.077	0.614	−0.585	1.276	0.952	0.711	0.532	1.089	0.899	0.725
Total New Mexico												
Dry	1.009	0.411	0.936	1.036	0.030	0.970	0.985	1.000	1.015	0.978	0.986	0.997
Productive	1.018	0.491	1.061	0.965	−0.032	1.024	1.008	0.992	0.976	1.016	1.006	0.995
Oil	1.016	0.552	1.092	0.958	−0.040	1.046	1.025	1.005	0.985	1.033	1.022	1.008
Gas	0.992	1.702	1.817	0.672	−0.586	1.220	0.910	0.679	0.507	1.035	0.848	0.676

Table B4 – Comparisons of Joint Association Survey Sample and Data as Published, 1959* (continued)

Data area and type of well	$\hat{I}_1^2/I_0^2$	S_1/S_0	K_1/K_0	α_1/α_0	$\alpha_1-\alpha_0$ ($\times 10^{-4}$ omitted)	H_1/H_0	MC_1/MC_0 5,000 ft.	MC_1/MC_0 10,000 ft.	MC_1/MC_0 15,000 ft.	Y_1/Y_0 5,000 ft.	Y_1/Y_0 10,000 ft.	Y_1/Y_0 15,000 ft.
North Dakota												
Dry	1.001	0.587	1.106	0.960	−0.087	1.060	1.015	0.972	0.930	1.035	1.002	0.965
Productive	1.002	0.966	1.362	0.889	−0.269	1.211	1.058	0.925	0.809	1.113	1.009	0.897
Oil	1.003	0.777	1.407	0.873	−0.304	1.229	1.055	0.906	0.779	1.123	1.002	0.877
Gas	Insufficient data											
Oklahoma												
Dry	1.001	0.289	0.942	1.016	0.037	0.957	0.975	0.993	1.012	0.967	0.981	0.998
Productive	1.003	0.554	1.030	0.989	−0.025	1.018	1.005	0.993	0.980	1.009	1.000	0.990
Oil	1.007	0.578	1.171	0.937	−0.140	1.098	1.024	0.954	0.890	1.069	0.999	0.938
Gas	1.007	0.627	0.974	1.008	0.016	0.981	0.989	0.997	1.005	0.986	0.991	0.998
East Texas												
Dry	1.012	0.220	1.256	0.938	−0.176	1.178	1.079	0.988	0.904	1.125	1.041	0.961
Productive	0.996	0.257	1.352	0.912	−0.261	1.233	1.082	0.950	0.834	1.135	1.024	0.909
Oil	0.990	0.930	2.124	0.786	−0.760	1.670	1.142	0.781	0.534	1.316	0.960	0.672
Gas	1.000	0.429	1.053	0.987	−0.035	1.039	1.021	1.003	0.986	1.028	1.014	0.984
Gulf Coast Texas												
Dry	0.994	0.815	1.167	0.961	−0.131	1.122	1.051	0.984	0.922	1.077	1.019	0.958
Productive	0.994	0.959	0.874	1.036	0.093	0.906	0.948	0.994	1.041	0.932	0.966	1.008
Oil	0.986	1.715	1.380	0.891	−0.319	1.230	1.049	0.894	0.762	1.112	0.982	0.849
Gas	0.999	0.540	0.974	1.007	0.017	0.981	0.989	0.998	1.006	0.986	0.992	1.000

Table B4 – Comparisons of Joint Association Survey Sample and Data as Published, 1959* (continued)

Data area and type of well	I_1^2/I_0^2	S_1/S_0	K_1/K_0	α_1/α_0	$\alpha_1-\alpha_0$ ($\times 10^{-4}$ omitted)	H_1/H_0	MC_1/MC_0 5,000 ft.	MC_1/MC_0 10,000 ft.	MC_1/MC_0 15,000 ft.	Y_1/Y_0 5,000 ft.	Y_1/Y_0 10,000 ft.	Y_1/Y_0 15,000 ft.
North Central Texas												
Dry	1.019	0.295	0.939	1.019	0.048	0.957	0.980	1.003	1.028	0.971	0.989	1.010
Productive	0.989	0.716	2.457	0.558	-0.994	1.371	0.834	0.508	0.309	1.032	0.727	0.486
Oil	0.992	0.397	1.140	0.950	-0.112	1.083	1.024	0.968	0.916	1.026	1.005	0.957
Gas	1.001	0.248	1.495	0.739	-0.319	1.105	0.942	0.803	0.685	1.013	0.916	0.819
Panhandle Texas												
Dry	1.077	0.122	1.490	0.742	-0.202	1.105	0.999	0.903	0.816	1.065	0.988	0.930
Productive	0.994	0.990	0.972	1.056	0.072	1.027	1.065	1.104	1.144	1.048	1.073	1.102
Oil	0.998	0.842	1.972	0.653	-0.502	1.289	1.003	0.780	0.607	1.122	0.955	0.794
Gas	1.002	0.244	0.574	1.447	0.490	0.830	1.061	1.355	1.731	0.946	1.118	1.345
Southwest Texas												
Dry	0.997	0.454	0.959	1.008	0.021	0.966	0.977	0.987	0.998	0.972	0.981	0.990
Productive	0.997	0.781	0.908	1.056	0.103	0.959	1.010	1.063	1.120	0.991	1.026	1.071
Oil	1.000	0.464	0.723	1.164	0.446	0.842	1.052	1.312	1.643	0.966	1.158	1.423
Gas	1.002	0.639	0.954	1.022	0.035	0.975	0.992	1.009	1.027	0.984	0.996	1.010
West Texas												
Dry	0.996	0.608	1.101	0.974	-0.060	1.072	1.040	1.010	0.980	1.054	1.030	1.004
Productive	0.999	0.876	0.921	1.051	0.072	0.968	1.004	1.041	1.078	0.987	1.013	1.041
Oil	1.000	0.728	0.974	1.017	0.023	0.991	1.003	1.014	1.026	0.998	1.005	1.014
Gas	1.006	0.263	1.295	0.888	-0.183	1.150	1.049	0.958	0.874	1.093	1.026	0.995

Table B4 — Comparisons of Joint Association Survey Sample and Data as Published, 1959* (continued)

Data area and type of well	i_1^2/i_0^2	S_1/S_0	K_1/K_0	α_1/α_0	$\alpha_1-\alpha_0$ ($\times 10^{-4}$ omitted)	H_1/H_0	MC_1/MC_0 5,000 ft.	MC_1/MC_0 10,000 ft.	MC_1/MC_0 15,000 ft.	Y_1/Y_0 5,000 ft.	Y_1/Y_0 10,000 ft.	Y_1/Y_0 15,000 ft.
Offshore Texas												
Dry	Insufficient data											
Productive	Insufficient data											
Oil	Insufficient data											
Gas	Insufficient data											
Total Texas												
Dry	0.999	0.499	0.865	1.045	0.108	0.904	0.954	1.008	1.063	0.934	0.974	1.022
Productive	0.999	0.587	1.128	0.942	−0.115	1.063	1.004	0.948	0.894	1.028	0.986	0.940
Oil	0.997	9.498	0.905	1.056	0.093	0.556	1.001	1.049	1.099	0.982	1.014	1.053
Gas	0.995	1.975	1.191	0.913	−0.186	1.088	0.991	0.903	0.823	1.030	0.962	0.889
Utah												
Dry	1.024	0.606	1.347	0.865	−0.196	1.165	1.056	0.958	0.868	1.015	1.035	0.961
Productive	1.034	0.719	1.061	0.965	−0.076	1.024	0.986	0.949	0.914	1.002	0.974	0.947
Oil	1.053	0.284	1.020	0.993	−0.015	1.013	1.005	0.998	0.990	1.008	1.002	0.996
Gas	Insufficient data[a]											
Wyoming												
Dry	1.020	0.559	1.646	0.852	−0.388	1.403	1.155	0.952	0.784	1.249	1.075	0.905
Productive	1.007	1.032	1.171	0.956	−0.074	1.120	1.079	1.040	1.002	1.096	1.068	1.037
Oil	1.012	0.747	1.317	0.809	−0.017	1.065	1.056	1.047	1.038	1.059	1.056	1.051
Gas	1.022	0.268	1.099	0.962	−0.051	1.058	1.031	1.005	0.980	1.043	1.026	1.005

[a]The fact that the sample data are insufficient while a good fit is obtained for the published data casts some doubt on the latter.

Table B5 – Comparisons of Cost Functions as Estimated from Joint Association
Survey as Published, 1955, 1956, 1959*

Data area, type of well, and years	K_1/K_0	α_1/α_0	$\alpha_1-\alpha_0$ ($\times 10^{-4}$ omitted)	H_1/H_0	MC_1/MC_0 5,000 ft.	MC_1/MC_0 10,000 ft.	MC_1/MC_0 15,000 ft.	Y_1/Y_0 5,000 ft.	Y_1/Y_0 10,000 ft.	Y_1/Y_0 15,000 ft.
Alabama										
Dry 55-56	0.896	0.928	− 0.207	0.808	0.729	0.658	0.593	0.761	0.705	0.646
56-59	3.606	0.452	− 1.048	1.596	0.945	0.560	0.329	1.191	0.842	0.565
55-59	3.163	0.408	− 1.255	1.290	0.689	0.368	0.196	0.906	0.594	0.365
Productive										
55-56	Insufficient data									
56-59	Insufficient data									
55-59	Insufficient data									
Appalachian										
Dry 55-56	7.703	0.192	− 2.412	1.481	0.443	0.133	0.040	0.742	0.318	0.121
56-59	0.341	4.058	1.757	1.382	3.327	8.009	19.28	2.326	4.070	7.971
55-59	2.623	0.781	− 0.655	2.048	1.478	1.065	0.767	1.726	1.295	0.959
Productive										
55-56	0.966	1.036	0.122	1.000	1.063	1.130	1.201	1.009	1.052	1.088
56-59	1.265	0.942	− 0.130	1.191	1.116	1.046	0.980	1.164	1.092	1.035
55-59	1.222	0.975	− 0.053	1.192	1.161	1.130	1.101	1.174	1.150	1.126
Arkansas										
Dry 55-56	2.079	0.671	− 0.847	1.395	0.913	0.598	0.392	1.087	0.793	0.559
56-59	1.804	0.723	− 0.479	1.304	1.026	0.808	0.636	1.142	0.957	0.794
55-59	3.753	0.485	− 1.326	1.820	0.938	0.483	0.249	1.242	0.759	0.444
Productive										
55-56	1.722	0.772	− 0.584	1.329	0.992	0.741	0.548	1.117	0.896	0.695
56-59	0.740	1.265	0.523	0.936	1.216	1.579	2.052	1.093	1.332	1.546
55-59	1.274	0.976	− 0.061	1.244	1.207	1.170	1.135	1.221	1.195	1.074

*Throughout this table, the subscript 1 denotes the estimate for the later year and the subscript 0 denotes the estimate for the earlier year, in any comparison. All comparisons are after correction to 1959 prices, where appropriate. See p. 93.

Table B5 – Comparisons of Cost Functions as Estimated from Joint Association
Survey as Published, 1955, 1956, 1959* (continued)

Data area, type of well, and years	K_1/K_0	α_1/α_0	$\alpha_1 - \alpha_0$ ($\times 10^{-4}$ omitted)	H_1/H_0	MC_1/MC_0 5,000 ft.	MC_1/MC_0 10,000 ft.	MC_1/MC_0 15,000 ft.	Y_1/Y_0 5,000 ft.	Y_1/Y_0 10,000 ft.	Y_1/Y_0 15,000 ft.
Onshore California										
Dry 55-56	Insufficient data									
56-59	0.872	0.984	− 0.018	0.858	0.850	0.843	0.835	0.854	0.849	0.843
55-59	Insufficient data									
Productive										
55-56	Insufficient data									
56-59	0.207	3.530	0.671	0.732	1.023	1.431	2.002	0.873	1.058	1.303
55-59	Insufficient data									
Offshore California										
Dry 55-56	Insufficient data									
56-59	Insufficient data									
55-59	Insufficient data									
Productive										
55-56	Insufficient data									
56-59	0.046	10.43	2.491	0.477	1.657	5.759	20.00	0.958	2.231	5.773
55-59	Insufficient data									
Total California										
Dry 55-56	1.025	0.901	− 0.109	0.923	0.874	0.828	0.784	0.897	0.867	0.833
56-59	0.722	1.101	0.100	0.794	0.835	0.877	0.922	0.814	0.842	0.875
55-59	0.740	0.992	− 0.009	0.734	0.731	0.727	0.724	0.731	0.730	0.729
Productive										
55-56	4.879	0.283	− 0.654	1.383	0.997	0.719	0.519	1.160	0.966	0.789
56-59	0.213	3.521	0.652	0.751	1.040	1.443	1.997	0.897	1.075	1.315
55-59	1.041	0.998	− 0.002	1.039	1.038	1.037	1.036	1.040	1.038	1.037

Data area, type of well, and years	K_1/K_0	α_1/α_0	$\alpha_1 - \alpha_0$ ($\times 10^{-4}$ omitted)	H_1/H_0	MC_1/MC_0 5,000 ft.	MC_1/MC_0 10,000 ft.	MC_1/MC_0 15,000 ft.	Y_1/Y_0 5,000 ft.	Y_1/Y_0 10,000 ft.	Y_1/Y_0 15,000 ft.
Colorado										
Dry 55-56	4.299	0.687	− 1.602	2.955	1.326	0.595	0.267	1.732	0.845	0.386
56-59	3.645	6.619	− 1.343	2.255	1.152	0.589	0.301	1.494	0.870	0.471
55-59	15.67	0.425	− 2.945	6.661	1.527	0.350	0.080	2.588	0.735	0.181
Productive										
55-56	2.125	0.763	− 0.683	1.621	1.152	0.819	0.582	1.318	1.011	0.745
56-59	0.836	1.232	0.508	1.030	1.328	1.712	2.207	1.201	1.460	1.829
55-59	1.777	0.939	− 0.175	1.669	1.530	1.402	1.284	1.583	1.476	1.363
Illinois										
Dry 55-56	indet.	indet.	0	0.973	0.973	0.973	0.973	0.973	0.973	0.973
56-59	0	∞	2.155	0.652	1.915	5.624	16.52	1.124	2.213	4.709
55-59	0	∞	2.155	0.635	1.865	5.477	16.09	1.064	2.081	4.426
Productive										
55-56	0	∞	3.200	0.480	2.377	11.78	58.32	1.145	3.407	11.63
56-59	2.035	0.642	− 1.145	1.307	0.737	0.416	0.235	0.924	0.589	0.351
55-59	0	∞	2.055	0.654	1.827	5.105	14.26	1.057	2.006	4.087
Indiana										
Dry 55-56	0.175	5.471	2.385	0.956	3.150	10.38	34.19	1.892	4.343	11.21
56-59	0.977	0.536	− 1.353	0.524	0.266	0.135	0.069	0.351	0.211	0.118
55-59	0.171	2.934	1.031	0.501	0.838	1.405	2.353	0.665	0.918	1.150
Productive										
55-56	1.699	0.745	− 1.034	1.265	0.755	0.450	0.268	0.909	0.585	0.356
56-59	∞	0	− 3.015	1.623	0.360	0.080	0.018	0.667	0.241	0.077
55-59	∞	0	− 4.049	2.126	0.280	0.037	0.005	0.607	0.142	0.028

Table B5 – Comparisons of Cost Functions as Estimated from Joint Association
Survey as Published, 1955, 1956, 1959* (continued)

Data area, type of well, and years	K_1/K_0	α_1/α_0	$\alpha_1-\alpha_0$ ($\times 10^{-4}$ omitted)	H_1/H_0	MC_1/MC_0 5,000 ft.	MC_1/MC_0 10,000 ft.	MC_1/MC_0 15,000 ft.	Y_1/Y_0 5,000 ft.	Y_1/Y_0 10,000 ft.	Y_1/Y_0 15,000 ft.
Kansas										
Dry 55-56	0.510	1.457	1.197	0.742	1.351	2.456	4.467	1.081	1.760	3.134
56-59	1.279	0.838	− 0.619	1.072	0.787	0.577	0.424	0.879	0.683	0.500
55-59	0.652	1.220	0.578	0.795	1.063	1.417	1.874	0.950	1.201	1.566
Productive										
55-56	0.575	1.457	− 0.785	0.838	1.241	1.836	2.719	1.054	1.421	1.974
56-59	1.405	0.779	− 0.554	1.094	0.829	0.629	0.477	0.929	0.754	0.593
55-59	0.808	1.134	0.231	0.916	1.016	1.156	1.297	0.980	1.071	1.170
Kentucky										
Dry 55-56	0.433	1.932	1.758	0.837	2.017	4.805	11.70	1.436	2.886	6.409
56-59	2.247	0.505	− 1.804	1.135	0.461	0.187	0.076	0.656	0.319	0.141
55-59	0.973	0.976	− 0.045	0.950	0.929	0.908	0.888	0.943	0.923	0.907
Productive										
55-56	indet.	indet.	0	1.065	1.065	1.065	1.065	1.065	1.065	1.065
56-59	0	∞	0.294	1.018	1.179	1.366	1.582	1.051	1.135	1.226
55-59	0	∞	0.294	1.084	1.255	1.455	1.685	1.082	1.167	1.261
Onshore Louisiana										
Dry 55-56	0.147	2.779	1.360	0.409	0.807	1.594	3.145	0.598	0.943	1.589
56-59	0.552	1.134	0.285	0.626	0.722	0.832	0.960	0.681	0.758	0.858
55-59	0.081	3.152	1.645	0.256	0.583	1.326	3.019	0.407	0.715	1.364
Productive										
55-56	0.582	1.286	0.484	0.748	0.953	1.214	1.546	0.860	1.025	1.255
56-59	1.360	0.847	− 0.333	1.152	0.976	0.826	0.699	1.045	0.925	0.805
55-59	0.791	1.089	0.151	0.862	0.930	1.003	1.080	0.900	0.949	1.009

Data area, type of well, and years	K_1/K_0	α_1/α_0	$\alpha_1 - \alpha_0$ ($\times 10^{-4}$ omitted)	H_1/H_0	MC_1/MC_0 5,000 ft.	MC_1/MC_0 10,000 ft.	MC_1/MC_0 15,000 ft.	Y_1/Y_0 5,000 ft.	Y_1/Y_0 10,000 ft.	Y_1/Y_0 15,000 ft.
Offshore Louisiana										
Dry 55-56	0.471	1.473	0.530	0.677	0.882	1.150·	1.499	0.786	0.938	1.155
56-59	1.203	0.788	− 0.370	0.948	0.788	0.655	0.544	0.854	0.753	0.648
55-59	0.567	1.132	0.160	0.641	0.694	0.752	0.815	0.671	0.707	0.749
Productive										
55-56	2.451	0.731	− 0.804	1.791	1.198	0.802	0.536	1.406	1.016	0.714
56-59	1.342	0.831	− 0.370	1.115	0.927	0.770	0.640	0.999	0.883	0.748
55-59	3.289	0.607	− 1.174	1.997	1.110	0.617	0.343	1.404	0.896	0.534
Total Louisiana										
Dry 55-56	0.449	1.402	0.638	0.629	0.865	1.190	1.638	0.757	0.953	1.243
56-59	0.735	1.014	0.031	0.746	0.758	0.769	0.781	0.753	0.761	0.771
55-59	0.330	1.421	0.669	0.469	0.655	0.916	1.279	0.569	0.725	0.958
Productive										
55-56	0.536	1.333	0.508	0.715	0.921	1.188	1.532	0.826	0.990	1.218
56-59	1.636	0.782	− 0.443	1.280	1.026	0.822	0.659	1.127	0.963	0.804
55-59	0.878	1.043	0.065	0.915	0.945	0.976	1.009	0.932	0.953	0.979
Michigan										
Dry 55-56	1.709	0.724	− 0.896	1.238	0.791	0.505	0.323	0.940	0.657	0.435
56-59	0.700	1.292	0.686	0.904	1.274	1.795	2.529	1.113	1.462	1.997
55-59	1.196	0.935	− 0.210	1.119	1.007	0.907	0.817	1.047	0.960	0.870
Productive										
55-56	5.726	0.254	− 1.298	1.454	0.760	0.396	0.207	1.021	0.678	0.428
56-59	0.284	3.671	1.179	1.042	1.878	3.388	6.107	1.435	2.075	3.136
55-59	1.626	0.932	− 0.118	1.516	1.429	1.348	1.271	1.464	1.407	1.341

171

Data area, type of well, and years	K_1/K_0	α_1/α_0	$\alpha_1 - \alpha_0$ ($\times 10^{-4}$ omitted)	H_1/H_0	MC_1/MC_0 5,000 ft.	MC_1/MC_0 10,000 ft.	MC_1/MC_0 15,000 ft.	Y_1/Y_0 5,000 ft.	Y_1/Y_0 10,000 ft.	Y_1/Y_0 15,000 ft.
Mississippi										
Dry 55-56	1.600	0.754	− 0.575	1.207	0.905	0.679	0.510	1.020	0.826	0.662
56-59	1.260	0.847	− 0.270	1.068	0.933	0.815	0.712	0.989	0.900	0.790
55-59	2.016	0.639	− 0.845	1.289	0.845	0.554	0.363	1.008	0.744	0.523
Productive										
55-56	2.890	0.620	− 0.885	1.791	1.151	0.739	0.475	1.386	1.010	0.700
56-59	0.287	1.667	0.961	0.479	0.774	1.252	2.024	0.633	0.895	1.335
55-59	0.830	1.033	0.077	0.857	0.890	0.926	0.962	0.877	0.904	0.934
Montana										
Dry 55-56	1.340	0.796	− 0.373	1.066	0.885	0.734	0.609	0.959	0.842	0.726
56-59	0.396	1.587	0.852	0.629	0.963	1.475	2.257	0.805	1.091	1.554
55-59	0.531	1.262	0.479	0.670	0.851	1.104	1.374	0.771	0.920	1.128
Productive										
55-56	1.178	0.931	− 0.139	1.097	1.023	0.955	0.890	1.043	0.993	0.939
56-59	7.008	0.213	− 1.280	1.492	0.787	0.415	0.219	1.056	0.710	0.461
55-59	8.256	0.196	− 1.419	1.619	0.796	0.392	0.193	1.101	0.706	0.433
Nebraska										
Dry 55-56	0.007	27.42	4.391	0.168	1.509	13.56	121.1	0.621	3.207	20.13
56-59	∞	0	− 4.557	5.154	0.528	0.054	0.006	1.286	0.239	0.036
55-59	∞	0	− 0.166	0.898	0.826	0.761	0.700	0.797	0.767	0.732
Productive										
55-56	0	∞	2.162	0.479	1.412	4.162	12.27	0.833	1.644	3.509
56-59	∞	0	− 2.162	2.213	0.751	0.255	0.086	1.178	0.596	0.279
55-59	indet.	indet.	0	1.059	1.059	1.059	1.059	1.059	1.059	1.059

Table B5 — Comparisons of Cost Functions as Estimated from Joint Association
Survey as Published, 1955, 1956, 1959* (continued)

Data area, type of well, and years	K_1/K_0	α_1/α_0	$\alpha_1 - \alpha_0$ ($\times 10^{-4}$ omitted)	H_1/H_0	MC_1/MC_0 5,000 ft.	MC_1/MC_0 10,000 ft.	MC_1/MC_0 15,000 ft.	Y_1/Y_0 5,000 ft.	Y_1/Y_0 10,000 ft.	Y_1/Y_0 15,000 ft.
Total New Mexico										
Dry 55-56	1.268	0.825	− 0.210	1.047	0.943	0.849	0.764	0.988	0.924	0.856
56-59	1.259	0.876	− 0.124	1.102	1.036	0.973	0.916	1.065	1.026	0.984
55-59	1.596	0.722	− 0.334	1.153	0.976	0.826	0.699	1.053	0.949	0.843
Productive										
55-56	0.993	1.015	0.014	1.008	1.015	1.022	1.029	1.012	1.016	1.021
56-59	1.056	0.929	− 0.068	0.981	0.948	0.917	0.886	0.963	0.944	0.922
55-59	1.048	0.943	− 0.054	0.988	0.962	0.936	0.911	0.974	0.958	0.941
North Dakota										
Dry 55-56	0.321	1.408	1.034	0.452	0.758	1.271	2.132	0.624	0.953	1.538
56-59	4.317	0.579	− 1.501	2.500	1.180	0.557	0.263	1.578	0.864	0.436
55-59	1.384	0.816	− 0.467	1.129	0.894	0.708	0.560	0.983	0.823	0.670
Productive										
55-56	2.124	0.692	− 0.813	1.469	0.978	0.651	0.434	1.140	0.851	0.598
56-59	0.804	1.179	0.327	0.948	1.116	1.314	1.548	1.052	1.176	1.349
55-59	1.708	0.816	− 0.486	1.393	1.093	0.857	0.672	1.200	1.000	0.807
Oklahoma										
Dry 55-56	1.276	0.889	− 0.236	1.134	1.008	0.896	0.796	1.059	0.970	0.880
56-59	0.565	1.270	0.507	0.718	0.925	1.192	1.536	0.834	1.008	1.253
55-59	0.722	1.129	0.271	0.815	0.933	1.069	1.224	0.883	0.978	1.102
Productive										
55-56	0.688	1.211	0.367	0.832	0.999	1.201	1.443	0.926	1.057	1.231
56-59	1.002	1.046	0.098	1.048	1.101	1.157	1.214	1.078	1.118	1.168
55-59	0.689	1.267	0.465	0.872	1.100	1.388	1.750	0.998	1.183	1.438

Table B5 — Comparisons of Cost Functions as Estimated from Joint Association
Survey as Published, 1955, 1956, 1959* (continued)

Data area, type of well, and years	K_1/K_0	α_1/α_0	$\alpha_1-\alpha_0$ ($\times 10^{-4}$ omitted)	H_1/H_0	MC_1/MC_0 5,000 ft.	MC_1/MC_0 10,000 ft.	MC_1/MC_0 15,000 ft.	Y_1/Y_0 5,000 ft.	Y_1/Y_0 10,000 ft.	Y_1/Y_0 15,000 ft.
East Texas										
Dry 55-56	1.880	0.729	− 0.799	1.370	0.919	0.616	0.413	1.077	0.798	0.551
56-59	0.567	1.241	0.517	0.701	0.908	1.176	1.522	0.817	0.986	1.253
55-59	1.061	0.905	− 0.281	0.960	0.834	0.725	0.630	0.881	0.786	0.692
Productive										
55-56	0.820	1.019	0.045	0.836	0.855	0.874	0.894	0.847	0.862	0.880
56-59	0.715	1.106	0.260	0.791	0.901	1.025	1.168	0.855	0.947	1.065
55-59	0.586	1.127	0.306	0.661	0.770	0.898	1.089	0.725	0.817	0.937
Gulf Coast Texas										
Dry 55-56	1.167	0.935	− 0.173	1.091	1.000	0.918	0.842	1.036	0.965	0.895
56-59	0.438	1.310	0.770	0.574	0.844	1.240	1.822	0.728	0.994	1.415
55-59	0.512	1.225	0.597	0.627	0.845	1.139	1.535	0.754	0.958	1.267
Productive										
55-56	1.192	0.899	− 0.233	1.071	0.953	0.848	0.755	0.999	0.916	0.828
56-59	0.533	1.276	0.571	0.680	0.904	1.204	1.602	0.807	1.003	1.290
55-59	0.636	1.147	0.338	0.729	0.863	1.022	1.210	0.806	0.920	1.069
North Central Texas										
Dry 55-56	1.226	0.844	− 0.425	1.035	0.837	0.677	0.547	0.910	0.772	0.639
56-59	0.788	1.112	0.257	0.876	0.996	1.133	1.288	0.946	1.045	1.170
55-59	0.967	0.938	− 0.168	0.907	0.834	0.767	0.705	0.862	0.807	0.747
Productive										
55-56	1.081	0.933	− 0.134	1.008	0.943	0.882	0.822	0.969	0.924	0.875
56-59	1.665	0.683	− 0.583	1.137	0.850	0.635	0.474	0.965	0.790	0.629
55-59	1.800	0.637	− 0.716	1.146	0.801	0.560	0.391	0.935	0.731	0.550

Table B5 — Comparisons of Cost Functions as Estimated from Joint Association
Survey as Published, 1955, 1956, 1959* (continued)

Data area, type of well, and years	K_1/K_0	α_1/α_0	$\alpha_1 \cdot \alpha_0$ ($\times 10^{-4}$ omitted)	H_1/H_0	MC_1/MC_0 5,000 ft.	MC_1/MC_0 10,000 ft.	MC_1/MC_0 15,000 ft.	Y_1/Y_0 5,000 ft.	Y_1/Y_0 10,000 ft.	Y_1/Y_0 15,000 ft.
Panhandle Texas										
Dry 55-56	1.566	0.804	− 0.412	1.259	1.025	0.834	0.679	1.113	0.964	0.812
56-59	3.877	0.344	− 1.109	1.336	0.768	0.441	0.253	1.006	0.692	0.467
55-59	6.086	0.276	− 1.521	1.682	0.786	0.368	0.172	1.120	0.668	0.379
Productive										
55-56	2.126	0.627	− 0.907	1.332	0.846	0.538	0.342	1.023	0.736	0.502
56-59	1.099	0.889	− 0.170	0.976	0.896	0.823	0.756	0.931	0.879	0.824
55-59	2.336	0.557	− 1.078	1.300	0.758	0.447	0.258	0.952	0.646	0.414
Southwest Texas										
Dry 55-56	0.944	1.028	0.073	0.971	1.007	1.045	1.083	0.990	1.018	1.051
56-59	0.770	1.047	0.123	0.806	0.857	0.911	0.969	0.837	0.879	0.930
55-59	0.728	1.077	0.196	0.784	0.865	0.954	1.052	0.829	0.895	0.979
Productive										
55-56	1.476	0.806	− 0.561	1.190	0.899	0.679	0.513	1.003	0.804	0.624
56-59	1.702	0.826	− 0.407	1.405	1.146	0.935	0.763	1.251	1.072	0.901
55-59	2.512	0.666	− 0.969	1.672	1.029	0.634	0.391	1.254	0.862	0.562
West Texas										
Dry 55-56	1.163	0.952	− 0.061	1.107	1.074	1.041	1.010	1.087	1.067	1.046
56-59	0.266	1.852	1.032	0.492	0.825	1.381	2.313	0.661	0.948	1.437
55-59	0.309	1.763	0.971	0.545	0.885	1.439	2.339	0.719	1.012	1.502
Productive										
55-56	0.973	1.031	0.035	1.003	1.020	1.039	1.057	1.012	1.023	1.036
56-59	0.637	1.314	0.356	0.838	1.001	1.196	1.429	0.924	1.040	1.188
55-59	0.715	1.355	0.390	0.969	1.178	1.431	1.739	0.935	1.064	1.231

Data area, type of well, and years	K_1/K_0	α_1/α_0	$\alpha_1-\alpha_0$ ($\times 10^{-4}$ omitted)	H_1/H_0	MC_1/MC_0 5,000 ft.	MC_1/MC_0 10,000 ft.	MC_1/MC_0 15,000 ft.	Y_1/Y_0 5,000 ft.	Y_1/Y_0 10,000 ft.	Y_1/Y_0 15,000 ft.
Offshore Texas										
Dry 55-56	1.933	0.618	− 0.418	1.194	0.969	0.786	0.638	1.073	0.940	0.816
56-59	Insufficient data									
55-59	Insufficient data									
Productive										
55-56	0.183	1.320	1.201	0.242	0.441	0.804	1.466	0.365	0.619	1.111
56-59	Insufficient data									
55-59	Insufficient data									
Total Texas										
Dry 55-56	0.999	1.006	0.014	1.005	1.012	1.019	1.026	1.009	1.014	1.021
56-59	0.748	1.086	0.197	0.813	0.897	0.990	1.092	0.862	0.933	1.014
55-59	0.749	1.093	0.212	0.817	0.913	1.010	1.124	0.870	0.944	1.035
Productive										
55-56	1.006	0.983	− 0.029	0.989	0.975	0.961	0.947	0.981	0.971	0.961
56-59	0.733	1.191	0.302	0.874	1.017	1.182	1.375	0.952	1.060	1.197
55-59	0.738	1.171	0.274	0.864	0.991	1.136	1.303	0.934	1.029	1.150
Utah										
Dry 55-56	2.435	0.616	− 0.440	1.500	1.204	0.966	C.775	1.331	1.163	1.001
56-59	0.295	1.775	0.547	0.523	0.687	0.904	1.188	0.607	0.719	0.869
55-59	0.718	1.093	0.107	0.785	0.828	0.874	0.922	0.808	0.836	0.870
Productive										
55-56	∞	0	− 1.867	2.076	0.816	0.321	0.126	1.212	0.685	0.363
56-59	0	∞	2.103	0.312	0.893	2.555	7.315	0.530	1.023	2.139
55-59	0.575	1.126	0.236	0.648	0.674	0.758	0.854	0.643	0.700	0.777

Table B5 — Comparisons of Cost Functions as Extimated from Joint Association
Survey as Published, 1955, 1956, 1959* (continued)

Data area, type of well, and years	K_1/K_0	α_1/α_0	$\alpha_1 - \alpha_0$ ($\times 10^{-4}$ omitted)	H_1/H_0	MC_1/MC_0 5,000 ft.	MC_1/MC_0 10,000 ft.	MC_1/MC_0 15,000 ft.	Y_1/Y_0 5,000 ft.	Y_1/Y_0 10,000 ft.	Y_1/Y_0 15,000 ft.
Wyoming										
Dry 55-56	0.846	1.196	0.319	1.012	1.187	1.392	1.633	1.107	1.241	1.414
56-59	0.639	1.150	0.293	0.735	0.851	0.985	1.141	0.801	0.893	1.012
55-59	0.540	1.376	0.612	0.743	1.009	1.370	1.861	0.887	1.108	1.432
Productive										
55-56	0.775	1.164	0.234	0.902	1.005	1.140	1.281	0.963	1.043	1.144
56-59	0.949	0.974	− 0.043	0.925	0.905	0.886	0.867	0.913	0.900	0.884
55-59	0.735	1.134	0.191	0.834	0.918	1.009	1.103	0.879	0.939	1.012

177

SUPPLY AND COSTS
IN THE U.S. PETROLEUM INDUSTRY:
Two Econometric Studies

BY FRANKLIN M. FISHER

designer:	Edward D. King
typesetter:	Monotype Composition Company
typefaces:	Baskerville
printer:	The Murray Printing Company
paper:	Bookshire Offset
binder:	The Murray Printing Company
printed cover:	John D. Lucas Printing Company

Printed and bound by CPI Group (UK) Ltd, Croydon, CR0 4YY

24/10/2024

01778539-0001